Telecommunications for Every Business

How to Select the Best Equipment for Your Enterprise

Robert D. Lewis

Bonus Books, Inc., Chicago

96 95 94 93 92 5 4 3 2 1

Library of Congress Catalog Card Number: 91-77016

International Standard Book Number: 0-929387-73-2

Bonus Books, Inc.
160 East Illinois Street
Chicago, Illinois 60611

Distributed by **abc Teletraining**
Geneva, IL

Printed in the United States of America

Composition by Point West, Inc., Carol Stream, Illinois

To the memories of my mother, Barbara Lewis, and brother, Michael Lewis. Both offered constant encouragement of whatever I happened to be pursuing at the time, regardless of what it was. I miss them both.

Contents

Acknowledgments

I'd like to thank everyone who had a hand, knowing or otherwise, in the preparation of this book. Especially, the exceptionally talented staff of the *Star Tribune's* Communications Systems Department, and of that group, Fonda Stewart-Calvert, Joe Arone, Bob Sellstedt, and Terri Crane made this book possible. I also owe a world of thanks to Mike Beaumont, Steve Eliason, and Chuck Ogden of AT&T, and Stella Rosenquist and Pat Walker of US West, for their ongoing efforts to educate me in how to be a better telecommunications customer.

To my wife, Ruth, who knows this subject better than I, thanks more than I can express for both education and encouragement. To Traci Taghon, my editor, and Aaron Cohodes, who owns Bonus Books and first suggested there might be a book in the strange way I choose to spend my days, thanks, both for your friendship and the opportunity.

Finally, I have to thank my father, Herschell Gordon Lewis, who gave me my start in Show Business and who shares at least a small bit of blame for introducing me to Aaron Cohodes in the first place. Thanks, Dad.

SECTION I

THE BASICS: DIAL TONE

CHAPTER ONE

Here's Your Advantage

HOW CAN A SMALL-TO-MEDIUM-SIZED business compete against the giants? Huge corporations should have an overwhelming edge—in resources, in economies of scale, and in corporate credibility.

In fact, if you run a business of 5,000 or fewer employees (even better if you have 500 or fewer), you have some tremendous advantages. The most important is your ability to fully understand your business. You don't think the CEO of Sears can understand the details of running department stores, credit card operations (Sears and Discover), insurance underwriting, and a stock brokerage, do you?

It's a miracle giant corporations can survive at all. They can for only one reason: Their construction resembles the brontosaurus, which had a second brain in its pelvis to supervise the activity back there.

You have another big advantage, although you may not realize it. You can take advantage of the same communications technologies the giants use, without having to manage the complexity. For example, many corporate giants maintain their own, private long-distance networks, just to avoid being buried in long-distance expenses for internal calls.

Your whole business is in one building—no nationwide private network needed. They have a whole highly-paid staff you don't need to hire and manage, let alone the expensive equipment you don't need to buy and maintain.

But where it counts, you can use the same sophisticated technologies the giants use, and without having the cost of that technology eat up your profits. Quite the opposite: **Everything described in this book can provide you with solid strategic and financial advantages—the kinds of benefits even the most conservative corporate controller would endorse.**

One caveat: You don't *have* to hire a staff to manage the technologies in this book. You may, as a business decision, *choose* to do so. Most of the technologies you'll read about in this book run themselves, but they can't change themselves to fit your changing business conditions, nor can they fix themselves when they break. You can contract with outside vendors for these services, or you can hire employees for the same purpose.

The Most Important Payoff Idea in this Book: You can contract with outside vendors for all kinds of support. You *can't* contract out overall responsibility. You must retain ownership of the basic technologies. This means keeping up-to-date on developments so you can plan your overall strategy for communications technologies. The companies that have tried to contract out basic responsibility—the industry term is "outsourcing"—are beginning to reverse that decision.

In this book you will find everything you **need** to know to effectively install, upgrade, or improve your telephone systems, local phone service, and long-distance service . . . and how to manage them.

I've written this book for the nontechnical business manager who has to make technical decisions.

No, change that. I've written this book for the nontechnical business manager who has to make *business decisions* regarding communications technology.

Some of what is in this book may be familiar to you. For example, in chapter 3 you'll read about the difference between PBXs, key systems, and Centrex/Centron telephone service. You very well may understand this already, but the next section provides detailed guidance in selecting the best product for your business.

This is a business book, not a technical book. If you want to understand how T1 signalling works, and the difference between it and ISDN protocols, look elsewhere. If you want to know how to decide which one **you** should buy, read on.

I've taken pains to avoid using jargon in this book. For example, I call the person who answers your central switchboard an "operator," not a console attendant. For that matter, I'll often call the console a switchboard, because "switchboard" is more familiar to you if you aren't in the industry. Similarly, you won't read anything in this book about "taking a phone off-hook." I'll say "pick up the phone."

The one disadvantage you'll find to this approach is that you won't get the hang of using the jargon, knowledge of which can be helpful when you talk to people who are in the telecommunications industry.

Read on. You are in the best of all possible situations—you can get nearly all of the benefits of modern communications technology, without having to endure the complexity.

Feel sorry for your larger competitors, who have to hire staffs of trained experts to obtain the same advantages.

But not too sorry.

CHAPTER TWO

Analyzing Your Requirements

NO MATTER WHAT communications technology you're considering, you can't avoid the process known as **analyzing requirements** (also known as figuring out what you need). You can find whole books on the subject, and you can spend oodles of money on consultants to help you.

Or, you can read this chapter.

You don't have to go through excruciating pain. In some respects, the process can approach enjoyability. (No, it won't actually *reach* enjoyability, but it can get close.) Your key to success: don't think of it as obligatory red tape. Understand it as the **most important success factor in your project.**

What you have to do is understand the problem you're trying to solve.

Here's another way of looking at it. By analyzing your requirements you take control of your own destiny, wrenching it away from technology gurus and salespeople who will inundate you with sales brochures describing electromagical features that will revolutionize your business.

(If you believe them, skip to page 643 where we talk about "Buying Your Share of the Brooklyn Bridge from the Author." The telephone really did revolutionize business. The personal computer is changing business now—chapter 11 tells you how—but revolutionary? Modern digital phone systems are revolutionary *technology*, but they won't revolutionize your business. A new technology, Custom Call Processing, may actually revolutionize your business. You'll find details on page 169.)

Payoff Idea: When you analyze your requirements, begin with realistic *business-oriented* goals. Depending on your current responsibilities, this may seem obvious.

Here is a line from a *Request for Proposal* (RFP) I reviewed not too many years ago: "The system must be able to handle multiple levels of software interrupt." This isn't a business requirement, and doesn't belong in any of *your* RFPs.

When you make your list of requirements, inspect each one and ask yourself if it specifies technology or states a business need.

To RFP or Not to RFP

You may choose to produce a formal RFP*, a less-formal *Request for Information* (RFI), or to work directly with salespeople in a more interactive process. This is your choice, or a matter of your company's policy. **Don't regard an RFP as an obligation, unless your company policy makes it so.** Use an RFP to explain what you need to several vendors while avoiding a lot of repetitious meetings. Also use it as a tool to make sure you have been fair to every bidder.

If your project will require a formal contract (as opposed to a simple warranty statement from the vendor) RFPs become more valuable. After you select a vendor, that vendor's response becomes the bulk of your contract. In other words, this is work you'll have to do anyway.

Whether you plan on creating an RFP or not, start out by creating a checklist of your requirements on paper. Put it in a logical, outline format. Now think about explaining your requirements to your vendors and keeping track of their answers in clear terms, and in a form that facilitates comparison. Finally, decide how much formal documentation you'll need to

*A sample RFP is provided on page 199.

explain the process later. This should tell you how formal an evaluation process you need to undertake.

If you're an expert, an RFP makes a great deal of sense. If you're just learning the subject, though, you'll find no faster way to gain expertise than to sit through several sales presentations. Once you've done so, you probably won't need an RFP.

Your Goal: No matter how you decide whom to buy from, you have two goals. First, and most important, you want to select the best product for your company. Second, you need to avoid playing favorites—intentionally or otherwise—among the competing vendors. The good news is, you can't accomplish one without also accomplishing the other.

In technobabble, "requirements" generally include REQUIREMENTS, Requirements, and requirements. Make sure *you* identify each of your requirements as critical, significant, or frill. (Don't ignore frills. A feature you don't really need, and that you have lived without for all these years, can still save you money or add value to your operation. Or, it can provide a better or more economical way of doing something.)

Also, ask an independent pair of eyes to review each of your requirements to ensure they describe *what* you need, not *how* your vendor should provide it. You will face no more difficult task in the world of technology than separating these two.

Here is a generic RFP for you to use as a starting point in creating your own. If you won't use an RFP, still review the outline to make sure you cover whatever items you need.

Something You Should Probably Do: Have your lawyers review your RFP, just in case.

Generic Request for Proposal

I. Table of Contents: Be a sport. Help the vendors navigate your RFP.

II. Purpose: What is this document for? Divide this section into...

   A. *What you want:* In general terms, what you're looking for. You're looking for a telephone system that can handle 1,000 telephones, or a personal computer network that can service 25 PCs now, but can grow to several hundred systems over the next few years.

Keep it to a paragraph or two. You'll go into detail later on.

B. *Who you are:* Describe your business and why you're interested in buying something. Give some detail about your operation, so your vendors can describe a solution to you, instead of a neutral description of technology. If you're buying a telephone system, vendors need to know whether you use telephones for general business only, or if you have some areas with special needs. If you're buying an Automated Call Distributor, vendors need to understand how your phone rooms operate now.

C. *Your goals:* What you hope to accomplish. Do you want to reduce your staffing? Do you want to provide better service to your customers? Of course you do—those are Mom and apple pie. Try something specific, like reducing the time customers spend on hold, or providing twenty-four-hour limited service through automation.

III. General Instructions: How to respond and what to expect. Give a deadline, the number of copies you want, a name and address to send them to, and a phone number in case anyone has a question. Also tell the vendors when to expect your decision. If you view any information in the RFP as confidential, tell the vendors here—don't expect them to figure it out osmotically.

Payoff Idea: *Tell* vendors how long their bids must be good for, rather than asking. If you'll need 3 months to make a decision and sign a contract, tell the vendors their bids must be good for 120 days. (No, not 90 days. You've never had an unexpected delay?)

Common Practice: Insist vendors explicitly note any deviations from your specifications in their responses.

Another Common Practice: Describe a response format, *if you think it will help you evaluate responses.* Don't do it "just because." Even if you don't prescribe an exact response format, remember to ask for a firm, fixed price and references.

Payoff Idea: Invite vendors to give more information than you've asked for. You may have missed something.

IV. Requirements: What you're looking for. Organize this section into subsections (General Requirements, Technical Specifications, Business Specifications) to make your own life easier when evaluating responses.

A. *General Requirements:* Such basics as whether you'll accept used equipment, custom-designed equipment or software, or vaporware (the industry term for promised-but-not-yet-available technology). Also use this section to describe how long you'll want what you're buying to last.

Very Important: Ask for detailed component information, in Item—Quantity—Unit Price—Total Price format. You'll probably want to explore the impact on cost of several slight changes in configuration, and going back to the vendor for each of these is a terrible waste of time.

Payoff Idea: Ask for a guarantee that the vendor will provide parts and service throughout the projected life of the system.

B. *Technical Specifications:* The meat. What you want the system to do.

1. *Reliability:* It is a rare business that won't list this as a critical requirement. You will find it tempting to engineer reliability in your RFP. Don't.

   If you ask, "Do you provide redundant power supplies," it is easy to answer "Yes." If you ask, "What factors have you engineered into your PBX to ensure reliability," you make sure the answer means something. More important, you don't take the risk of missing important or innovative approaches. As an example, one vendor may offer redundant power supplies, but a competitor may provide independent power supplies on each circuit card. By asking the specific question, you make it easier to understand distinctions among competing products.

2. *Stage in Life Cycle:* How long has this product been on the market, and how long do this vendor's products typically stay on the market? You want, of course, to make sure you can get parts and support as long as you are using the system. A critical re-

quirement. (You asked in General Requirements, but what will you do in five years if they change their minds? Get some reassurance.)

As a general rule, the best time to buy technology is eighteen months after its introduction to the market. Each technology has its own rules, though, and you'll find specifics in the following sections.

3. *Environment:* Any special considerations for housing what you're buying. In what range of temperatures does it operate reliably? Does it require additional air conditioning? Raised flooring? Will humidity be a problem?

   How much space does it occupy? How much additional space do you need around it for clearance? How much does it weigh?

   How much power does it use? Is an Uninterruptable Power Supply (UPS) built in? If not, what kind of UPS will you need, and at what price will the vendor provide it? (Be careful when comparing responses. UPSs come in several flavors, so as you compare each vendor's inflated price with what you can find in the discount catalogs, make sure you're making a fair comparison.

4. *Features:* The fun stuff. The crux of the biscuit. The most difficult section to write, because...

   You should describe what you want to accomplish, not how to accomplish it—you should provide your organization's business requirements, not the vendor's technical specifications. On the other hand, it's easier to simply tell vendors you want call conferencing than it is to explain that employees often need to talk with more than one individual at a time over the telephone.

   Don't be obsessive. Make sure every feature you list maps to real business needs, but if you want a specific feature, don't be afraid to ask for it.

   Caveat: Whenever you list specific features, make sure you've offered vendors the chance to offer equivalent services in different ways.

5. *System Design and Architecture:* The engineering spe-

cifics. Ask about system design *only when you care about it*. For any system that has to hook up to equipment you already own, you care. For most others, you don't. Don't ask "just because." Anything you ask, vendors must respond to, and you'll have to read their responses.

6. *Administration:* What it takes to keep things going on a day-to-day basis. Don't ask for detail in your RFP—there's rarely any point to it. Ask how much training is required, and ask for an estimate of how many FTEs (Full-time Equivalent Employees) would normally be needed to administer a company of your size.

   If it makes sense for your organization, ask about having your vendor contract to administer your system for you. For smaller organizations (and many larger ones) this is a good idea.

C. *Business Specifications*

1. *Vendor Experience:* How many systems has the vendor installed, and especially, how many of the model you will be buying has the vendor installed? How many are within your metropolitan area, or a 100 mile radius?

   You may choose to work with a younger, growing company, especially if they give you some financial incentives. There's nothing wrong with that. You only have a problem if you find you're working with an inexperienced company after you've signed your contract.

2. *Warranty and Maintenance Terms:* Warranties generally last either ninety days or a year, but ask anyway. Maintenance after that generally runs from 10 to 15 percent of the purchase price, but again, don't assume.

   If you're open for business outside of normal business hours, ask about terms for after-hour or twenty-four-hour service coverage.

   If you're hoping for consequential damages (repayment for lost business and so forth) when things go wrong, forget it. You won't find a single company in the technology business that doesn't exclude consequential damages in its contracts.

Payoff Idea: Ask how many service technicians are employed within fifty miles of your location and how many systems they are responsible for. Restrict answers to the specific product you are installing. A vendor may have a huge staff, but only three technicians trained on the product you're buying.

Don't accept any handwaving. The phrase, "...but it's basically the same as this other product ..." should lead you to ask, "Then why am I buying this one, if the two are the same?"

Also ask about remote maintenance. Many systems provide the ability to do nearly anything by modem. This helps vendors by letting them do more from their central location. It also helps you, because your vendor can respond faster.

Don't consider remote maintenance to be a panacea. You'll need it when your system breaks, and when your system breaks, all bets are off. Plan on remote maintenance handling at least 70 percent of your problems.

3. *Payment Terms and Conditions:* For most items, expect to pay a third up front, a third on installation, and a third on acceptance. For simpler, more generic items such as PCs, you pays your money and you takes your chances.
4. *Training:* Everyone provides training, and it's always comprehensive, and provided by professionals who *really care.* Of course they do.

   Ask how many training sessions your vendors will provide, the maximum class size, and the nature of the training materials they provide.

   Also ask about administrator training. Expect your vendor to provide training for as many individuals as you will need to administer the system, but at least two, unless you will contract with your vendor for system administration. (If training isn't local, you will have to pay travel expenses yourself. Be reasonable. If you're buying a dozen PCs you can't expect much free training. A half-million-dollar telephone system is another story.)

5. *References:* References should be restricted to companies that use the same product you are investigating. You need, of course, at least three, preferably local, references, including a contact for each.

Payoff Idea: Form a task force composed of your internal experts. If you're buying a new telephone system, for example, include your main operator and several receptionists from around the organization, and have them review your RFP and attend all sales presentations. Of all the people in your company, they will have the most insight into your telephone requirements. Take advantage of their expertise.

Evaluating Responses

Evaluating responses to RFPs can be a thoroughly intimidating business. For every point in your RFP, each vendor will either have written a lengthy response, a long, bulleted list, or the phrase, "Understood and complied with," a phrase you'll come to treasure.

Everyone has different ways of evaluating RFP responses. Here's one that may work for you.

1. Read through each response, without any formal grading or scaling. After each one, write down your overall impressions, positive and negative. When you finish the last one, decide whether you see a clear winner, or any vendors who have clearly disqualified themselves.
2. Read through each response a second time. Grade each answer on a 5-point scale—5 is a perfect fit, 0 is completely unacceptable. Use prepared scoring sheets for this purpose. When you are done, multiply the score on each point by 1 (for frills), 2 (for significant features), or 3 (for critical items). Don't be locked into a 3-point scale, though—you may have a few points that warrant a 4- or 5-point multiplier.
3. See how well your numerical scores match your impressions. If they match, you're done. If they don't, do some skullwork to determine what it is about your impressions that don't jibe with the more detailed analysis. You may decide your

weighting scheme is off. Or you may reevaluate your impressions.

Don't do a solo evaluation, and do produce a written recommendation. This is an area where more than one pair of eyes is essential.

Finally, don't be afraid to make a decision. Especially in a group process, the urge to "run through it one more time, just to be sure" can be nearly overwhelming. Resist it, and encourage the group to resist it. Keep asking yourself if you will ever have any more information to help you with your decision. If you do need more information, go get it. Usually, you'll find that a group delays a decision in the hopes that something new will occur to someone.

It won't.

Once you have made your decision, check references. Ask about their general level of satisfaction, and anything they recommend you be aware of. Then ask about each point you have listed as critical.

It is rare that you'll encounter a really bad reference. Your vendor simply won't tell you about any installations that were truly disastrous. When you go through the detailed list you are more likely to come up with cautionary notes.

Nobody will be perfect, of course, so don't expect it. Your goal is to ensure the vendor will do the job.

Try to keep your process as informal as you can. The more you maintain a relaxed, casual atmosphere, the more you will learn.

CHAPTER THREE

Choosing a Telephone System

IMAGINE BUSINESS BEFORE THE TELEPHONE, if you can. The amount of time spent communicating through couriers, carrier pigeons, Western Union and so forth must have been staggering.

Compared to that, even a phone system seems economical.

If you work in a modern office, you already know a great deal about telephone systems. You've had little choice. Your knowledge may not be systematic, but as you read this chapter you will find less new material than you might expect.

You'll find a lot of basics in what follows. Everyone knows something about the subject, but most people will have gaps, and no two people will have the same gaps.

Business Summary

Useful Life:	7 years
Cash Cost:	\$500–\$750 per seat
Amortized Cost:	\$250–\$500 per seat per year, including local and long-distance service
Dedicated Staff:	Optional
Space:	< 25 telephones (approx) wall-mount in wiring closets
	> 25 telephones require about 1 ft^2 per 10 telephones
Temperature:	< 300 telephones (approx)—normal office
	> 300 telephones—depends on system
Electrical:	Uninterruptable Power Supply needed. Power requirements vary widely.
Justification:	You're not in business without a phone system.

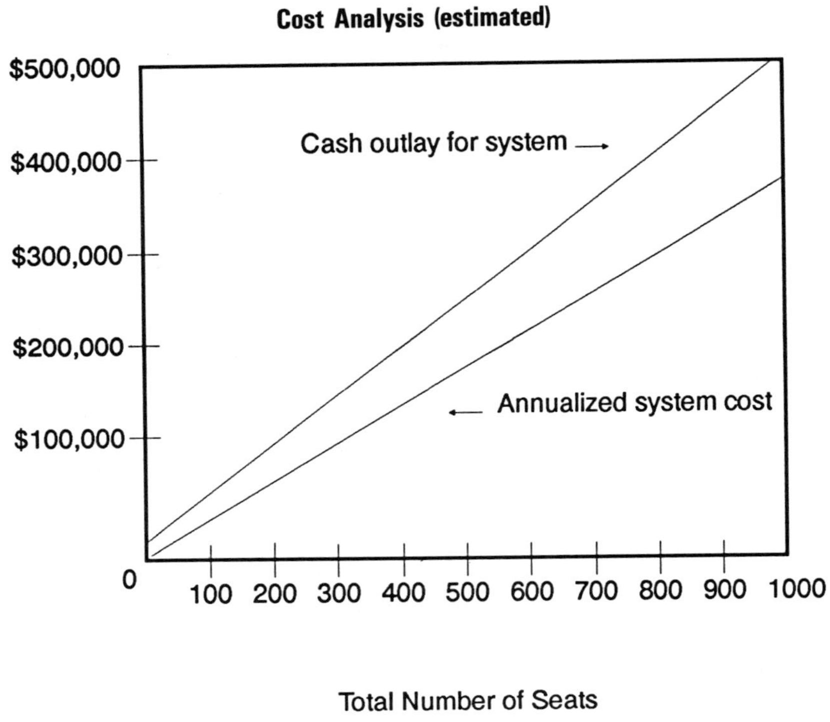

SIGNALLING

Telephones can communicate in three ways: Analog signalling, DSS digital signalling, and ISDN digital signalling.

Analog communication is easy to understand. An electrical voltage mimicks the sound waves you produce when you speak. Analog lines are relatively cheap, they work reliably, but they are prone to be noisy.

DSS — Digital Switched Service — represents sound waves as numbers. You don't need to understand exactly how, since all you do is buy products that take care of it all. DSS can be electrical or fiber optic. It is more expensive than analog service, but is effectively noise-free (if the other end of the line also has digital service).

ISDN—Integrated Services Digital Network—is a new standard for digital telephony. Despite tremendous industry hoopla, it provides only two features of real significance, and neither of those will affect you for a few years. These are the easy ability to piggyback both voice and data transmission over the same connection, and Automatic Number Identification (ANI)—the ability to determine the phone number of the individual calling you. Other than this, ISDN has the same advantages and disadvantages for you as DSS.

SWITCHING

Telephone systems can route calls using analog or digital switching technology, but analog and digital in this context have little to do with analog and digital signalling.

An analog phone system uses a physical switch—like a light switch—to connect incoming calls to phones. The system's hardware contains everything it needs in order to connect each call to the proper phone.

A digital phone system routes calls by converting all information into discrete packets of digital data, then routing each individual data packet to the appropriate destination. Digital phone systems are computers, and are controlled by software. This makes them tremendously flexible, and makes them susceptible to software bugs as well.

Some phone systems are hybrid analog/digital devices. In these, the switching logic is controlled through software, but signals remain analog and are routed using software-controlled physical switches.

ALTERNATIVES

You can put phones in your business in three and a half different ways: Phones provided by your local telephone company (Centrex and Centron), Key Systems, Private Branch Exchanges (PBXs), and Hybrids.

Centrex and Centron aren't all that different from having a home telephone. The local phone company runs wires to every desk, puts phones at the ends, and everyone can send and receive telephone calls. The phone company adds some bells and whistles so you can put people on hold, transfer calls, hold conference calls and so forth, but many of your employees will never bother to learn these frills.

Centrex/Centron service has one key distinguishing characteristic: You don't actually own anything, and don't have to do any system administration unless you want to.

The difference between Centrex and Centron is this: Centrex service comes out of older, analog central office switching equipment, while Centron service comes out of newer, digital central office switches. As a result, Centrex provides fewer features (although it may have every feature *you* need), has poorer sound quality (analog signals are prone to be noisy), and (you'll like this one) generally costs a lot less.

With Centron service you get digital signalling and every feature you can imagine or read about below.

Terms you'll need when you talk to their sales force

RBOC: Pronounced "AR-boc," it stands for Regional Bell Operating Company, the seven phone companies created when the feds broke up AT&T. Depending on where you live, you'll deal with Nynex, Bell Atlantic, Southern Bell, Ameritech, USWest, Southwestern Bell, or PacTel.

CO: Central Office. This isn't the main business location, despite the sound of it. In phone company jargon, a central office is a computerized telephone switching system to which private phones or phone systems attach. It may seem odd, but the phone company may have several different COs in one building. Even in the same room.

POTS: Plain Old Telephone Service. Proof that in technical fields, no excuse whatever is needed to create acronyms.

Key Systems used to be simple phone systems you could install in your office. Every line coming into the business appeared on every phone, and you picked the line you wanted by pressing the corresponding button.

Remember those? They were simple, easy to understand, and never broke down.

Things have improved.

Modern key systems are fully digital, automatically find an open line when you want to dial out, and support nearly every feature you can imagine. (If your imagination doesn't run in this direction, the next section provides a list of the more important features you can ask for in a phone system.)

In a typical key system, each phone has access to a specific group of outside lines.

Key systems are the smaller of the two types of private phone systems you can own. Depending on which of the fifty or so systems on the market you choose, they can support from 2 (but why bother?) to about 300 phones.

At the 300-phone end of the key system range, the distinction between key systems and PBXs (below) nearly vanishes. The next section will describe what distinctions remain.

Terms you'll need when you talk to their sales force

Interconnect:	In the trade, any company that sells and installs private phone systems.
KSU:	Key Service Unit. The box that runs a key system.
KSU-less:	Just what you think. A key system that doesn't need a central box to run it.
Voice Terminal:	A telephone. Just as Human Resource Professionals receive higher wages than members of Personnel Departments, so Voice Terminals cost more than telephones.
Analog/Digital	Internally, the KSU operates the same way a computer does, but all communication to the phones uses analog signalling.
Fully Digital:	The KSU uses digital logic, and communication to the phones uses digital signalling.
UPS:	Uninterruptable Power Supply. You can buy books on just this subject if you're masochistic, but with a key system or PBX, you need a UPS.

PBX stands for Private Branch Exchange. You need to know this about as much as you need to know that *Laser* stands for "Light Amplification by Stimulated Emission of Radiation."

When you own a PBX, you become, in effect, your own central office. You no longer use business lines—you use trunks. (When I first heard the word "trunk," I envisioned a thick, meaty cable that handled lots and lots of volts. The reality is less imposing. A trunk is just another pair of wires, electrically no different than a business line.)

PBXs are computerized switchboards. Your local RBOC sells you trunks. The number of trunks you buy is equal to the number of simultaneous outside calls you need to handle. The PBX connects each outside call to the appropriate inside line.

Except on the used equipment market, you'll have a hard time finding analog PBXs. The leading products (that is, every product you should consider) are digital.

The good news is modern PBXs make a staggering array of features available to you. (Whether anyone in your company actually uses them is an entirely different matter, of course.)

The bad news is that digital PBXs are driven by software, and no software is completely bug-free. The good news about the bad news is that most of this stuff is pretty good. You will always hear horror stories, but they are the rare exception in this business.

Put it a different way: Your PBX *will crash*. . .occasionally. . .but it won't go down hard and stay down for long. Typically, if it goes down due to a bug, you'll be without your phones for five minutes or so.

Terms you'll need when you talk to their sales force

Switch:	Another term for a PBX.
Switch Room:	Oh, come on. Guess.
Trunk:	A generic term for the type of phone line that goes from the central office to a PBX.
DID:	Direct Inward Dial. Your ability to assign specific telephone numbers (as opposed to extensions) to a particular phone in your office.
DTMF:	Dual Tone MultiFrequency. The sounds your touchtone phone makes, except that AT&T has registered Touchtone® as a trademark.
Autoattendant:	The computerized voice that asks you to enter the extension if you know it, or, "For Sales, Press 1. . ."
Console Attendant:	Operator.

Hybrid Systems have some characteristics of key systems and some characteristics of PBXs. They are designed to service midsized businesses—between about 100 and 500 phones—and while they are very hard to differentiate from PBXs, they may be a perfect fit for you. Hybrid systems are evolving very rapidly; a few even accept digital central office service, once the exclusive province of the PBX.

Many executives probably long for the "good old days" of telephony, when everyone had a four-button telephone. You didn't transfer calls, you put the caller on hold and yelled, "Hey Fred, pick up line three." You didn't conference calls, you yelled, "Hey Fred, pick up line three," and stayed on line three yourself, too.

Ah, why can't life still be that simple?

Answer: You now have 500 employees, and to have enough buttons on your phones, each one would be the size of a desk blotter.

And so, we have the modern, complicated telephone system.

Having used telephones since childhood and throughout your business career, you will know most of what follows instinctively. Probably, it will clarify some points for you, and include others you may have missed.

Review chapter 2 for generic requirements. Here are the issues specific to telephone systems:

Requirements

I. Technical Specifications
  A. *Station Features:* The features available to your users. Everyone knows at least three of these. Every PBX has all of these. Specify them anyhow, to understand *how* they are provided. If you plan to purchase a key system make sure you specify the ones you need. Here are the most important, in no particular order:
    1. *Call Transfer:* Transfer a call to someone else.
    2. *Call Pickup:* Pickup someone else's call. Part of administration is defining call pickup groups.
    3. *Call Hold:* Annoy your caller.
    4. *Conference Calls:* Talk to more than one individual at a time. Nearly all systems place some restrictions on the number of participants and/or the mix of inside and outside callers. Make sure you ask for a description of the limits imposed by each system you are investigating. (If you expect to exceed them on an occasional basis, you can make use of outside teleconferencing services instead.)
    5. *Coverage:* Have a designated phone answer your calls if your line is busy or you don't answer it. You may need the ability to establish different coverage paths for these two conditions. Also make sure you have a forward-to-cover option—equivalent to telling your secretary, "Hold all my calls."

Another option is directing internal calls to a different coverage point than external calls. In a sales environment, for example, you may want internal calls to go to voice mail, but external calls to go to another sales representative.

6. *Automatic Redial:* If you get a busy signal, monitor the recipient's line. When it is clear, the phone system automatically dials you both. (If you plan to install voice mail, this option has little value.)
7. *Call Forward:* Redirect any calls to a different phone. (This is similar to coverage except that you designate a different line each time you use this feature.)
8. *Call Parking:* Transfer a call to a different phone and have it wait until you get there to pick it up.
9. *Speed Dialing:* Store phone numbers on programmable keys (see B, Multiple Phone Types) to quickly dial important or frequently-used numbers.
10. *Priority Calling:* A way for high-priority callers (i.e. the president of your company) to signal you that you have an important call (e.g. from one of them).
11. *Intercom:* Talk to someone else inside your office. Significant only if you are planning to buy a key system, and every key system known has this feature.
12. *Message Waiting:* Have your phone signal you that you have a message. May be signalled by the operator, or by your voice mail system (see chapter 7 for more details).
13. *Multiple Calls:* Allow more than one call per extension (ask how many). The system should allow any combination of incoming and outbound calls, and should place no restrictions on conferencing among them.

B. *Multiple Phone Types:* Keep in mind that your vendor will refer to telephones as voice terminals. Chances are you will need more than one type of telephone. For each of the following features and special telephones you need, determine (a) if you can get it; (b) what phones it is available for; (c) how easy it is to learn and use (especially on the basic set you'll give to 90 percent of your employees).

Here are most of the features and special types of telephones you are likely to look for:

1. *Hold Button:* The basic telephone may have a designated hold button or a function key that can be programmed as a hold button, or it may require pressing a key combination to place people on hold. Or, the basic set may lack this feature altogether.
2. *Message Waiting:* The basic telephone may provide a message waiting light, or it may provide a special tone when you pick up the handset (the light is better, of course). Look for a way for employees to turn off their own message waiting lights.
3. *Access to other Special Features:* Determine how conveniently each type of telephone is able to use the various station features listed above.
4. *LCD display:* One- or two-line displays that provide the caller's name and extension.
5. *Speakerphone:* Conversation through a speaker and built-in microphone. Look for the phone's ability to operate in *full-duplex* mode—to allow both people to talk at once. This has become the rule rather than the exception, but don't assume it.
6. *Hands-free Speakerphone:* A speakerphone that can go off-hook (that is, you can answer it) without physically picking up the handset.
7. *Headset and Cordless Headset:* For employees who spend large amounts of time on the telephone, and for whom speakerphones are inappropriate, headsets are vital.
8. *Data Interface:* A standard connector into which you can connect a PC to control the phone and use your phone system's data-handling abilities. Look for the type of connector (should be DB-25) and the interface (should be RS-232-C configured DCE). In the companion to this book, *Communications Systems for Every Business,* same publisher, you'll learn superior alternatives to using a PBX for data-handling.

Payoff Idea: Instead of listing your requirements (and suffering the inevitable migraine that will result) ask all respondents to list the options they provide. Then figure out which ones you want, on the theory that (1) there is no point in asking for features nobody provides; and (2) the manufacturers may provide a feature you can use but didn't think to ask for.

Caveat Emptor Department: Your vendor will emphasize the snazzy features on their expensive egophones. You should instead scrutinize the basic telephone that 90 percent of your employees will get. Ask if it has a real hold button or just an assignable function key. Also ask if employees can turn off their own message waiting lights. Remember, the phone's design determines the system's ease of use. Your phone system really can make your employees more productive, if they learn to use its features. If their phones are an obstacle, you'll get less out of your investment than you could.

C. *System Console:* What a modern telephone system provides instead of the old-fashioned switchboard. Depending on how big an operation you have and your expansion plans, look for the ability to handle multiple consoles.

Payoff Idea: As a part of your review process, have your central operators review the console designs of all vendors responding to your RFP. They're in a better position than anyone else to evaluate your requirements, and by involving them in the selection, they'll feel more ownership in the results.

D. *Automatic Route Selection (ARS):* Commonplace on PBXs and rare in key systems, this is the ability to automatically select the lowest-cost service for long-distance calls placed through your phone system. A great feature, but not critical since there are other ways of managing long-distance costs (see page 84). Rate it as significant or frill, depending on your business. If you are large enough that it is critical, you will end up buying a PBX that has it anyway.

Caveat Emptor Department: Remember that if your long-distance rates change or you change or add long-distance vendors, you will have to update your ARS tables. This is obvious now, but easy to forget later.

E. *DISA (Direct Inbound System Access):* Most PBXs now have this feature and it's becoming more common on key systems. This feature allows your employees to call your system from a remote site and then call an outside number. The application: Access to reduced-cost long-distance service.

There are two basic scenarios. In the first, an employee calls in through your WATS line to reach a local number. For example, you are attending an out-of-town seminar and need to call a local client (local to your business, that is). Instead of calling direct, you call your own business through its inbound WATS line, then dial your client. The call costs you at WATS rates instead of your hotel's inflated long-distance charge.

The second scenario is the opposite. For example, you may need to call a supplier long-distance from a customer site. Instead of placing a credit card call, you call your own business, then use its outbound WATS lines to make the long-distance call.

Payoff Idea: Change the access password to this feature frequently, or just use credit cards and don't take the risk. An increasingly common form of telephone abuse is for outsiders to try phone systems at random until they find one set to the default access code. Then they rack up huge long-distance bills at your expense.

F. *Data Handling:* Fully-digital PBXs and key systems can transmit voice and data calls simultaneously and independently over the same pair of wires. A snazzy feature. Rate it a frill, regardless of your business requirements, because you can take care of your data communications requirements in many other ways.

Remember, the manufacturer of your PBX has complete control over what phones you can plug in, so voice compatibility isn't an issue. Data transmission is more finicky and idiosyncratic, so you can expect setup problems for any device more complicated than a PC. (You can expect setup problems with PCs, too.)

At one time, the idea of using PBXs to handle both voice and data communication seemed attractive, and many PBX manufacturers spent a great deal of time and energy building data-handling features into their systems. Local Area Networks (LANs) have made the subject largely irrelevant. You're better off. If you seek more details on LANs, please consult the companion to this book, *Communications Systems for Every Business.*

If you have more than one location, you may find great advantages from the ability to handle data. Ask what provisions the vendor makes for emulating a "leased line," a special circuit provided by telephone companies which links locations without one of them having to dial the other. If you have more than one location of any size, you should probably hire a consultant to talk about the best ways of linking the two.

G. *Computer Interfaces:* Controlling your phone system with a computer, and sending call information to a computer. The main control feature is dialing. Call information may include ANI for inbound calls and call status (ringing, busy, etc.) for outbound calls. Pay close attention to price—the feature may be useful, but not *that* useful. Also look for the ability to interface with computers other than PCs. You may want to feed ANI to the mainframe as part of a phone room operation, for example.

H. *Analog Switching:* The ability of fully-digital systems to accommodate old-fashioned analog devices. Probably not a problem but check anyway. You'll need analog service to connect answering machines, facsimile machines, and modems. Rate it critical if you use any of the above devices.

Caveat Emptor Department: If you are about to move from analog to digital technology and rely heavily on answering machines, be prepared. You won't be able to retrieve calls your machine has picked up. Why? Instead of your answering machine sharing a line with your phone, it will have a separate analog line. When you don't answer your phone, it will forward the call to your answering machine.

Good News Department: If you are talking on the phone and a second call comes in, your answering machine will be able to take the call.

I. *ISDN:* You probably aren't ready to plunge into ISDN service, given its newness and cost (check with your phone company though). You still may want to keep your options open. If so, make sure the system you are considering can be upgraded to ISDN.

As of this writing, there are no key systems that have the option of using ISDN. This, of course, could change tomorrow.

J. *Call Detail Reporting:* A computer interface that provides information about every call, also called SMDR for Station Message Detail Reporting. You can either connect the CDR port to a printer (which will waste a huge amount of paper for no good purpose) or you can connect it to a PC-based system to track costs and charge them back.

Look for the system's ability to filter this information. In an age of metered service, you'll probably need to charge back both local and long-distance calls, but you almost certainly won't want to keep track of internal conversations.

Availability of CDR is significant if your organization requires chargebacks. (If you need chargebacks, you almost certainly are large enough to require a PBX, and all PBXs sport this feature.) Evaluate it on the flexibility of filtering, data provided, and on what it can do for you without an attached CDR system.

II. Business Specifications

A. *Warranty and Maintenance Terms:* Ask, but expect the usual one year warranty. Maintenance after that will generally run about 10 to 12 percent of the purchase price.

Expect remote maintenance as an option on all but the smaller key systems.

B. *Payment Terms and Conditions:* Typical terms are a third up front, a third on installation, and a third on acceptance.

Payoff Idea: Form a task force composed of your main operator and several receptionists from around your company, and have them review your RFP and attend all sales presentations. Of all the people in your company, they will have the most insight into your telephone requirements. Take advantage of their expertise.

Another Payoff Idea: Make sure you include Centrex and Centron service as options when you send out your RFP. When you compare responses, remember that with either of these options you avoid the costs of an Uninterruptable Power Supply, the cost of climate-controlled space needed for a PBX (or just the space needed for a key system), and possibly the cost of paying one or more employees to administer a phone system.

Evaluating Responses: Review the advice in chapter 2. Of what you find there, by far the most important advice is to **have your task force of operators and receptionists participate in the evaluation.**

Installation and Cutover

The good news is you won't have to do any of the difficult stuff when you install a new telephone system. Your vendor will handle everything, from project planning and tracking, to generating the bill of materials and ordering components, to pulling cable, to. . . you name it, you're paying for it.

The bad news is you will have to do the really tedious work, and it is all still your problem anyway.

If you've ever been through a telephone system installation, little of what follows will be new. The tips are still of value.

What to Expect from Your Vendor

Your relationship with your vendor may be the most curious association in business. From the time you sign the contract until the last telephone is cut over to the new system, you will work for your vendor as much as the vendor will work for you.

Your vendor will assign a project manager, and will ask you to assign (or be) a project manager. Yes, there will be two PMs. Your vendor will defer to you in public, but this is simply a courtesy. For the most part, you work for the vendor. Be grateful. You don't want this job. From here on, "project manager" refers to the vendor project manager.

How much of the following advice you need depends on the size of your installation. It is written for a system with at least 100 lines. Use your common sense in scaling things back, if your installation is smaller.

The First Meeting: This is a getting-acquainted session. Your vendor's whole project team should be present. So should yours.

Payoff Idea: Invite your account representative from the local phone company, and perhaps representatives from your long-distance vendors. The main purpose of this meeting is to ensure that everyone knows whom to call, and is comfortable with roles and responsibilities.

You may be tempted to take on a coordinator's role, and ask that all communications happen through you. **Resist this impulse.** The amount of coordination needed for successful installation is formidable. The last thing you need is self-imposed communications barriers.

A point of etiquette: You are officially in charge. This means that you should kick off the meeting, say how pleased you are to be working with everyone, and ask all the people in the room to introduce themselves and explain their roles.

Then, turn the meeting over to the project manager. At this meeting, expect generalities, and global descriptions of responsibilities. Schedule weekly meetings to discuss progress, problems, scheduled tasks and so forth. You will know what these are because the first thing you should receive from your vendor is...

The *project plan* is one of the reasons you don't want to be the real project manager. The PM will provide you with a detailed project plan, including specific tasks, deadlines, and responsibilities. Put differently, the project plan will tell you what needs to be done, when it will happen, and who will do it.

The one perspective from which you really are in charge is this: You monitor the plan, make sure you understand it, and make sure your vendor meets the deadlines. If any are missed, meet *immediately* with the project manager to determine how the vendor will make up the time.

A Warning Sign: If the PM tells you a project plan is superfluous, bells should go off in your head. Even for an eight-line key system, where you may not have a formal project manager and everything is less formal, your vendor should tell you who is responsible for the installation, what the critical events are, and when they should occur.

If your vendor tells you not to worry because the company has done a hundred of these, start worrying. If you don't have a list of events and dates, neither does your vendor. Most vendors know what needs to be done, but they don't know when.

Insist on a project plan.

You Don't Need to Hear this, but. . . Department: Don't let your ego get in the way. If you don't understand something, ask. The people on the vendor team don't expect you to have their level of knowledge, and they all want you to. The more you understand the technology they are installing, the easier their lives will be.

What Is Expected of You

When you signed the contract, you dumped most of your headaches onto the project manager. You have, however, retained the most thankless jobs in the installation:

1. Estimating trunking requirements (you can get a lot of help with this)
2. Determining station requirements
3. Defining staff parameters
4. Resolving special situations
5. Coordinating training
6. Listing critical phones

The discipline of defining trunking requirements is known as *traffic engineering*. Your vendors can provide a lot of support, but the final decision is yours. How many central office trunks (for a PBX) or 1FB lines (for a key system) will you need to support your call volume? You'll find guidelines in the section on Traffic Management on page 61.

For the rest of your responsibilities, be smart. Insist that each organization within your company appoint a liaison for you to work with. Make all decisions their decisions. Meet with them once a week, and invite a vendor representative to attend the meetings (the training coordinator is a good choice).

You don't have to be especially bright to determine *station requirements.* You do have to be patient and persistent. All you have to do is decide who receives each type of phone and the features programmed into each. Remember, a PBX can supply many more special features than a telephone has special buttons, so you'll have to decide what to include on the phone. (A point of diplomacy: Call them "voice terminals" in vendor discussions, but "telephones" when talking to users.) If your company has 500 employees with phones, you have 500 decisions to make.

See why you need your liaisons?

Ask your vendor liaison to describe the characteristics of each type of voice terminal to your liaison group, along with typical decision criteria for each. Your vendor should provide glossies on each phone for all your liaisons, so they can go back and meet with their groups.

After the presentation, provide your liaisons with the cost of each type of phone, so they can make reasonable business decisions. As with the rest of the installation, count on your vendor to do most of the dirty work. You'll have a lot of coordination to take care of, and you don't want to be too far away from the action. The more ego phones you buy, the more you'll spend, and you'll be amazed at how fast this can turn into real money.

While you're determining station requirements, initiate a search for analog lines (assuming you're installing a digital system). You will need to know where analog circuits will be needed in your new configuration. Primarily, you will need them to support answering machines, fax machines, and modems.

Provide your liaisons with a list of all employees in their area at the beginning of the process. Take the time to set up an employee database *before* the installation begins. Then, providing an employee list is easy. It's simply a printout of your database, sorted by organization.

Everything depends on the size of your installation, but

for a 500-phone system plan on taking no more than two months to determine station requirements. Six weeks is better, and for no reason—except procrastination or vacation schedules—should the process take more than a month. Your problem, of course, is that you're dealing with volunteer labor.

Record all station decisions in your database, or in your notebook if you've decided to use pencil and paper techniques.

Staff parameters sounds intimidating. . . the sort of thing a consultant might come up with. Sorry. Defining staff parameters isn't technically intimidating, but it is a daunting task.

What you're trying to do is define:

1. Call pickup groups
2. Coverage paths
3. Feature availability
4. Long-distance authorization (if you plan to restrict long-distance access)

If you're replacing an existing system you can make your life easier by skipping this step and going with the patterns you already have. But. . . why was it you decided to put in the new system?

Record all staff parameters in your database or notebook.

Payoff Idea: Use your new phone system as an opportunity to rethink your way of doing business. Your suppliers and customers probably contact you more by telephone than any other form of communication. You're putting in a shiny new system with new capabilities. Take a fresh look at how you want your phones answered: Everything from what people say (phone etiquette is a valid training issue) to what happens when they aren't at their desks.

This is a lot of work. It also can be the difference between customer complaints and improved customer relations, so, work with your customer satisfaction group to make some philosophical decisions, then with your liaisons to make practical decisions.

While you are deciding on station requirements and user parameters, use your liaisons as an investigative team to locate potential *special situations*—situations in which you can't duplicate an employee's current phone environment with the new system. Special situations are especially prevalent when you convert from analog to digital phones.

Let's handle the most common examples, beginning with answering machines.

Answering Machines: In an analog environment, you plug answering machines in the wall and phones into answering machines. The answering machine listens to your phone and after some number of rings, it picks up the call and takes care of it. If you're in your office and decide you want to speak to that caller, you can get the call back since your phone and the answering machine share the same line.

That doesn't work in a digital environment, since your phones use digital signalling and answering machines are purely analog. You have to run a separate analog line, and define it as the coverage path. If the phone isn't answered, or if it is busy, the phone system will transfer the call to the answering machine, which is now dialed at a separate number.

This is a good news/bad news situation. The good news is answering machines now handle calls when your line is busy, as well as when you can't answer. The bad news is you can't use an answering machine to screen your calls.

Tape Recorders: Here's one that can make you reach for the aspirin bottle. If you have 500 employees you undoubtedly have one who is accustomed to tape recording calls. Attaching a tape recorder to an analog phone is pretty simple—most consumer electronics stores sell a gimmick that goes from the back of the phone to a cassette recorder's line input jack.

In a digital world you need a different strategy. Fortunately, the solution is pretty simple. The same consumer electronics stores that sold the other interface sell a "Y" connector that plugs into the handset jack of any phone. This takes care of the tape recorder, since the handset signal is analog (it is converted to digital signalling inside the phone). Try moving the existing interface to this new position before buying a new one. It may work. (This book avoids most legal issues, including the issue of notifying callers when you tape them. If you have an employee who tapes calls, look into the relevant statutes for your state.)

Fax Machines: Wherever you have a fax machine, install an analog line.

Modems: Wherever you have a modem, install an analog line. Most digital phone systems provide facilities for digital modem emulators and outside modem pools.

Payoff Idea: Before going ahead with this approach, look at the option of establishing a modem pool through a Local Area Network instead. Economics will probably favor the LAN approach, and LAN-based solutions will be more robust. You'll find more information on LANs and LAN-based modem pools in the companion to this book.

Call Routers: A recent innovation that may cause you some minor grief is the automated call router—a device that can detect whether an incoming call comes from a telephone, fax machine, or modem, and route it to the appropriate device. These allow all three devices to share a single phone line, which is a nice economy for an office that receives faxes and modem transmissions infrequently.

You're better off without these if you're installing a digital PBX or key system. Install a separate analog line for each destination device.

Shortly before you cut over to your new system, make sure everyone in the company attends a training session on the new phones. Keep close track of who fails to attend, since these are the people who feel they are too important to waste the time. If all else fails, you may have to provide them with one-on-one training, especially for those who really are that important.

Your vendor will provide training as part of your contract. **Make sure the trainers know about any service changes special to your installation.** For example, you may change your long-distance access code, or eliminate it altogether. Make sure your employees learn about it in their training class.

If you have a seven day a week operation, the project manager will ask you for a list of *critical phones*—phones you need in service immediately.

If you have a Monday through Friday operation your life will be easier—the employees will leave the old phone system on Friday, and when they return on Monday their new phones will be in place. Many businesses never close. For these businesses, some phones must return to service as fast as humanly possible. These are your critical phones. Your liaisons can identify them for you.

Phased Cutovers: Don't even think about it. If you are a big enough business to need a phased cut, you are way beyond the scope of this book. Do it and get it done. Otherwise your cut will hang around your neck like an albatross for the rest of your days.

Other Issues

Phone Number Changes: In some cases your new phone system, or simple business growth, may necessitate a change in phone numbers. This happens most often if you decide to change from analog trunks to DS-1 or ISDN central office trunks, or if you decide to unburden your switchboard by changing to DID service.

You won't find any magical formula in this book to make a phone number change any easier. What you will find is a checklist, to save you the trouble of compiling your own:

1. If you're getting DID service, have your Marketing Department decide which phones get the numbers ending in 00, and any other easy-to-remember numbers (examples, 3333, 9191, 1234). These are a precious commodity, and you don't want responsibility for their allocation.
2. Don't miss the deadlines for White Pages printing, and for all Yellow Pages you advertise in.
3. Print new business cards and stationery.
4. Print new business forms (you'll find at least a dozen forms that need changing, maybe hundreds). Ask your liaisons to help compile the list.
5. Notify your customers of the change a week before you cut over. Any earlier and you'll just confuse them, unless you *Remote Call Forward* the new number to the old one (see below). Then, start phasing in the new number a month before your cut.
6. Notify your employees of the change early and often. Repeat the information during their training sessions.

Remote call forwarding means the phone company automatically forwards calls dialed from one number to another. Before your cut, your customers will reach you at the old number if they dial the new one. **Do this. It allows you to publicize your new number early.**

Deciding whether to remote call forward after your cut-over is a more difficult decision, and depends on the details of what your customers expect from you. The service isn't free, so you'll want to be judicious.

Here are the most reasonable options.

1. Don't do it at all. Rely on the phone company's number change message ("The number you have dialed has been changed. . .") and your customers' basic good sense. This has the advantage of economy and also means you get it over with.
2. Remote call forward until the phone books with your new number have been distributed. This avoids time-wasting arguments between your customers and operators (". . . but the phone book says your number is. . .").
3. Remote call forward all numbers until new phone books are out, then continue to forward just your central number(s) until call volume is low enough that you are comfortable terminating the service. If your business is a fixture in the community, you may opt for this approach.

Rewiring: Some vendors include a sum for rewiring in their bids. Don't argue too much with this. Just because you already have wire to every office doesn't mean it will work. The demands of digital telephony are such that your old wire, if it was installed for an analog system, may not work.

In *Computer Systems for Every Business* you'll find information on Local Area Networks (LANs). There, you will learn about using telephone wire for Ethernet LANs. The marketing hype for this form of Ethernet, called 10BaseT, describes fabulous economies from its ability to use spare wires from the phone cable you've already installed.

Payoff Idea: If your vendor is pulling new wire for phones, have them pull data-grade cable to each office at the same time, and leave it hanging in the wiring closet. Despite the hype, you can't use spare wire pairs from your telephone cable for Ethernet LANs, and you'll save a ton of labor expense by pulling both at the same time.

Uninterruptable Power Supply: During the installation there will come a time when your vendor wants to turn on the new system and "burn it in" (this means turn it on and see if any components will fail immediately). The project manager may ask for a UPS for the new system.

This should be your vendor's expense, since you don't own the new system until you accept it. If you have an existing UPS that can handle the load, be a sport. Otherwise, unless the cost is low enough that you don't care, don't view this as your problem.

Absolutely Fundamental: Your phone system *must* have a UPS, to tide you over brief power outages, protect your system from voltage surges, and to allow an orderly system shutdown in case of a longer power outage.

911 Emergency Service: In environments where you dial 9 for an outside line, you need some way of providing your employees with an emergency number.

Many companies use an internal emergency number that connects to Security. If you are big enough to have a Security Desk this is probably your best bet.

You also can dial 9-911.

Whatever you do, instruct your employees early and often on what to do in an emergency. Remember, you may be the one to have a heart attack.You're installing the phone system, aren't you?

What Your Vendor Is Doing

Your vendor is plenty busy preparing your switch room for the new system, checking the wiring, programming the new switch based on the station requirements and user parameters you've supplied, arguing with their shipping department over when your phones will arrive, working with the phone company and your long-distance carriers...you don't need to know all of the headaches. Just be grateful they aren't your headaches.

The project plan is your guide. Despite the hands-off approach this book recommends, you do want to be familiar with the project plan. Read it until you understand it.

Why? If you're responsible for the new phone system, you won't sleep easily unless you understand the project plan. You can rest knowing the project manager has everything under control, because you'll know if everything is happening when it's supposed to.

Otherwise, the PM may tell you everything is fine, and you'll never know. And sometimes, everything isn't fine. Your only managerial responsibility is to detect problems early, and to take the necessary steps to correct the situation. If you've never had the pleasure of managing a contractor before, what follows may prove helpful. If you have, there's a good chance you've been through this before.

No installation follows the plan perfectly. Some things will inevitably miss their target dates. When that happens, ask for a revised schedule and an assessment of the impact the delay has caused by your next weekly update meeting.

Most delays are unimportant. If the phones are a week or two late, it doesn't matter very much.

If the programming is late, or the system itself didn't ship on schedule, your installation could be seriously jeopardized. Make sure you know the cause of all delays.

If you are faced with a serious hold-up, you need the answer to only one question: Is your project manager at fault, or is your vendor at fault?

While legally this is the same thing, right now you're trying to salvage the situation. If your PM appears incompetent (it happens more often than it should), don't be shy about asking for a replacement. If the installation goes badly, your neck is on the chopping block too.

More often, your PM will be the victim of poor business practices on the part of his employer—your vendor. If this appears to be the case, you have a key role to play as the project manager's advocate with his management.

The relationship between a PM and his management is often adversarial. He is competing with other PMs to secure the resources needed to make your installation successful. If you have to meet with vendor management, make it clear that in your view the project manager isn't the problem.

The Cutover

You have two, possibly three roles to play during the cut. First and foremost, you are responsible for keeping other employees from getting underfoot. Every company has at least one person with an undiscovered critical requirement that must be satisfied in the next five minutes or the business will have to file for Chapter Eleven.

You have to get that person away from the installation team.

Your second responsibility is to be a gofer. There are never enough warm bodies around. Check to make sure the critical phones are in service. Help "ring out" the phones (that is, make sure the right number is assigned to each phone). Double check the executive phones to make sure they're programmed correctly, have the right coverage paths, and so forth. Get coffee.

Also, check the special situations you uncovered earlier, to make sure they work the way they're supposed to.

Your final responsibility is to provide coffee and donuts in the morning, fried chicken at lunch, and pizza if the cut is especially long and runs over the dinner hour—unless your sales representative assumes this responsibility (the commissions on PBX sales aren't tiny).

In many respects, a telephone installation is a small miracle. Thousands of tiny details come together and result in a working system over the span of a few hours. While your part is not that large, you can still take a lot of pride in a successful cutover. It really couldn't happen without you.

The First Day of Business

The first business day after your cut is a lot like a hangover. After months of planning, you're suddenly back to business as usual except that:

- Nobody's phone works (translation: they've forgotten their training)
- The president can't get a call back from his answering machine (translation: all your patient explanations went in one ear and out the other)
- Somebody can't record his calls (he forgot to tell anyone of this requirement)

- A programmer's modem won't dial out (nobody knew there was a modem there in the first place)
- And so forth

In a cut of any size, your vendor will set up a Help Desk and provide floorwalkers for the first day or two, to handle any minor mistakes or changes that are needed. After that, it's back to managing a phone system.

This is the oldest cliché in the business, but there's no harm in repeating it: Nobody thinks about the telephone until it doesn't work (or doesn't work the way it used to). You're more likely to hear criticism about the 1 percent that works wrong than the 99 percent that are perfect.

Don't take it to heart. You did a good job.

CHAPTER FOUR

Managing Your New Phone System

FIRST, A FEW WORDS on facilities management in general: *Do as little as possible.* Without much effort, you can find enough work to keep as many people busy as you want to hire, and you can find justification for all of it.

Not keeping records can be a disaster. Too much recordkeeping can turn your productive technologies into black holes—you pour money in, and nothing comes out.

Knowing that none or too much recordkeeping is incorrect is not the same as knowing what amount is correct for your business. You may end up hiring a consultant or contractor to help you plan. *Caveat Emptor.* A consultant or contractor is not paid to describe the simplest possible approach to management. Even worse, many of them are trying to create permanent positions for themselves.

Whether you opt to develop a plan yourself with the help of this book or hire a consultant, here's a touchstone of sanity which is obvious and self-evident once stated.

Only record information you can use.

Put a different way, any information you record is information you'll have to maintain through all adds, changes, and deletions. Even a few of these can be a lot of work.

Only large systems require some form of recordkeeping.

If your company is a six person operation, don't bother. **Rules of Thumb Department:** If you can't walk into one wiring closet and immediately figure everything out, do some level of system management. Read on, and scale to fit your system.

The key to managing a phone system is **never lose track.** Once your records get out of sync, getting them back in is a monumental task.

To manage your system you need a database, either on paper or in a computer. If you want to computerize, you may want to consider one of the many telephone management packages currently for sale, or, if you are more adventurous, you may want to roll your own.

You may want to contract system management with your vendor. Depending on the size of your organization and how big a technical staff you want to keep, this can work out very well. If you choose to contract with your vendor, review what follows to ensure you both mean the same thing by system management.

To manage your phone system you have to manage:

- The PBX or key system itself
- Outside trunks or lines
- Users
- Telephones
- Cable
- Phone company charges
- Departmental chargebacks (possibly)
- Traffic

Managing a PBX or Key System

Modern phone systems are special purpose, modular computers. They consist of one or more cabinets. In each cabinet are shelves, in each shelf are cards (which go into slots), and each card has one or more ports. In some systems, any card can go anywhere. Other systems have different types of slots, none of which can handle all types of card.

Your job is to keep track of each port, and what it is attached to. It isn't difficult. Tedious, yes. Difficult, no.

Why You Care: This may seem obvious. Still, it's always a good idea to validate the reason for keeping track of something, to avoid the "just in case" syndrome.

You have two reasons for keeping track of ports in your phone system. The first is to help you or your vendor diagnose problems. If a phone has a problem, you'll usually need to know which port it's connected to. The alternative is for two technicians to visit wiring closets, one tugging on wires in one, the other watching them move in the other.

The second reason you want to keep track is to help you with adds, moves, and deletions. When you add a phone you need to know if you have any unused ports to plug it into. When you move a phone you'll often do it by wiring the new location into the old port (yes, Virginia, this is a case where fiddling with the hardware is often easier than making the change in software). Again, you need to know the port.

Keeping Track on Paper

Create a notebook, and tab it into sections based on cabinets. Then fill it with ledger pages as in Figure 1.

When you fill out the ledger pages, start a fresh page for each *Shelf #*, and leave several blank lines below the entries for each *Card #*. This way you'll be able to enter a number of changes before you have to rewrite a page.

Have these pages printed on a heavy, smooth stock—one that will allow repeated erasures without eroding. The first four columns won't change very often, and you'll find it a lot easier to handle changes by erasing and rewriting the *Extension, Type, User Name,* and *Location* than by lining out the entry and creating a new one elsewhere. Plus, you can keep your entries in order.

Keeping Track on Computer

Personal computers can keep track of this sort of detail without breaking a sweat. You're probably better off if you use one. Don't, however, underestimate the task. It isn't for the faint of heart.

Phone System Ledger Cabinet #_______ Shelf #_______

Card #	Type	Port	Cable #	Ext	Type	User Name	Location

Figure 1.
Sample Phone System Ledger

You can buy packages designed to manage telephone systems. If you buy a Call Detail Reporting System, you'll probably get telephone management for free.

If you want to set up your own system, you'll find guidance in Appendix A.

Managing Outside Trunks or Lines

Chapter 5 covers many of the issues surrounding outside trunks and lines in more detail. You don't have much to do in the way of managing these items. The phone company installs and maintains them.

You do need to keep track of the numbers assigned to each by the phone company in case you have a problem, and you have to keep track of how you use each one. The form in Figure 2 may be of value for this.

In the form, fill in your phone company's identification number for each trunk or line in the column labeled *Trunk/ Line #*. *Type* could be 1FB (for a key system), inbound analog, outbound analog, DID, analog two-way, DSS, PRI, or whatever.

Fill in the name of the phone company providing the line under *Carrier.* You do this because you may have trunks coming in from more than one carrier... see chapter 4 for more information on this subject.

Some PBXs let you reserve groups of trunks for special uses. You may, for example, identify a group of priority users who get their own reserved trunks. The last column, *Usage Notes,* gives you a place to keep track of this information. See Appendix A for suggestions on how to computerize these records.

Managing Users

When managing users on paper know what information you need to keep track of, in what order, and who really understands your end-users' needs. You're about to discover why a computerized system is vastly superior to a paper one for an installation of any size.

Before we go into recordkeeping...

Trunk/Line Ledger Page ________

Trunk/Line #	Type	Carrier	Usage Notes

Figure 2.
Sample Trunk/Line Ledger

Remember your departmental liaisons? The individuals who did so much work for you during the installation? Maintain these relationships forever. You'll need the support these people can provide. You can use them as filters to help you define departmental requirements and procedures consistently. You can use them as intermediaries, so that someone else can explain to an irritable department head that a system can't be set up in two conflicting ways at the same time.

Hold monthly meetings to go over any issues that have come up, and to introduce any new features your vendor may be installing. Make membership in this group a point of status.

Something Else You Don't Need to Be Told: Don't make any decisions, or system changes, without first informing this group. The liaisons are your lifeline to the company, and your way of explaining to management later that everyone hashed out the issues before you made the decision.

Keeping Track

For a typical installation, you need to know the name, location, phone extension, other phone extensions (called "appearances") for users with multiline phones, type of telephone, coverage path, call pickup group, and "class of service" (the way the phone system defines what features are available) for each user.

This isn't an exact list. Each PBX and key system is different. Set up your records based on the specific requirements of your system.

Why Do You Care: Good question. For ongoing system management, you don't really need more than name, location and extension, and even location is as much to help you print labels occasionally as for any management purpose. (After all, if you need someone's location you can always call him or her). You can find the coverage path, call pickup group, and class of service in the PBX management software itself, so you don't really need this information in your database.

You need to collect the other information while planning your installation and this database is the best place to store your planning information. Keeping the information up to date isn't much work, and you'll find uses for it once you have it in a database.

Keeping Track on Paper

Create a notebook. For a PBX, fill the notebook with prenumbered ledger pages based on the range of extensions available to you (see Figure 3). By prenumbering you'll have an easy way of finding available extensions when you have to add new users. Include several blank ledger pages at the back for special extensions you may want to use that are outside the normal range, and for phones that share extensions as described below.

You want this list in order of telephone extension because you normally have to publish an internal directory anyway, so you'll have a list sorted by user name.

If you are managing a key system, don't prenumber the records, since you'll have many phones with identical extensions (and, for that matter, appearances). Otherwise the recordkeeping is the same.

To fill out the form: *User Name* and *Location* are obvious. *Type* means the type of telephone. *Cover #* is the extension of the coverage path—where calls go when the phone is busy or isn't answered. *Group* is the call pickup group, if there is one. *Class* is the class of service.

The second-line entries *Cable #* and *Closet* refer to the number assigned to the cable running from the phone jack to the wiring closet, and the wiring closet itself. Your phone system vendor will assign cable numbers and can explain their numbering scheme. You can number or identify your wiring closets however seems appropriate. If they have room numbers, that is ideal.

App 2 through *App 6* are other appearances—other phone numbers that appear on the phone if it is a multiline phone.

This system works so long as each individual has his or her own telephone and extension. Two situations cause exceptions: More than one person can share a phone, and more than one phone can use the same extension (in a key system this is universal).

If more than one person shares a phone, handling the situation is easy in a pencil and paper system. Instead of writing in a user's name, write in the words, "See List 1" and insert

Telephone Extension Records: From *#5051* to *#5063* Page ______							
Ext	User Name/ Cable #	Location/ Closet	Type/ App 2	Cover/ App 3	Group/ App 4	Class/ App 5	App 6
5051							
5052							
5053							
5054							
5055							
5056							
5057							
5058							
5059							
5060							
5061							
5062							
5063							

Figure 3.
Telephone Extension Records

a list of individuals sharing that phone in the back of the Extension Record.

If more than one phone shares an extension, you can handle it in a similar fashion. Instead of writing in a user's name, enter the notation "See page 900" (or whatever). The page number refers to an Extension Record page you place at the end of the main group. This page is not prenumbered. On it you enter records for phones that share the same extension. Appendix A includes a description of a user database.

Managing a Telephone Inventory

You can manage any inventory in two ways: By quantity or by individual unit. In other words, you may just want to know how many telephones you have, or you may want to keep track of each individual telephone by its serial number. (Keeping track of each individual telephone is pointless.)

Why Do You Care: Telephones are a physical asset. You don't need to worry about employees taking them home, since digital telephones won't work in their homes (publicize this fact).

You do need to keep track of how many spares you have on hand for providing new employees with telephones. You also should make sure your vendor's technicians keep track of any of your phones they use to swap out telephones that need repairs. It's easy to become sloppy.

Keeping Track on Paper

In the back of the notebook you use for tracking users, keep a ledger sheet (see Figure 4). Each time you install a telephone or take one out of service, make an entry.

Figure 5 provides some sample entries to demonstrate the technique. In the example, 550 model 7401D digital telephones arrived on February 12, 1990, during the installation phase of a new PBX installation. These are logged as an *Add* transaction, and are added to the *Stock* inventory.

March 1 was the cutover date, so 495 phones were taken out of stock and installed. The system administrator (you) entered an *Install* transaction of 495 phones, added 495 to the in-

Telephone Inventory Ledger for Model ________

	Transactions				Inventories			
Date	Add	Install	Pulled	Repair	Install	Stock	Repair	Total

Figure 4.
Telephone Inventory Ledger

Telephone Inventory Ledger for Model *7401D*

	Transactions				Inventories			
Date	Add	Install	Pulled	Repair	Install	Stock	Repair	Total
02/12/90	550					550		550
03/01/90		495			495	55		550
03/15/90		12			507	43		550
04/03/90		1		1	507	42	1	550
04/17/90				-1	507	43		550
05/20/90			1		506	44		550

Figure 5.

stall inventory, and reduced the stock inventory by the same amount.

On March 15, a group of twelve contract programmers began work and needed phones. The system administrator entered another install transaction, for twelve phones, and adjusted the appropriate inventory levels.

On April 3, a phone failed and needed repair. A replacement phone was taken from stock. The system administrator entered both an install and a repair transaction. The install inventory remained the same, but the stock inventory was reduced, and the repair inventory increased.

On April 17, the broken phone came back from repair. A repair transaction of –1 and appropriate inventory adjustments were entered.

Finally, on May 20, an executive decided she wanted a fancier phone. The system administrator pulled her 7401D phone and added it to stock, replacing it with a 7506B. (The install transaction for the 7506B phone is not shown.)

If you want to keep track on computer, refer to Appendix A.

Managing Cable

The world is filled with horror stories regarding companies that lost track of their cabling. Since nobody knows which wire pairs are good and which ones aren't, the company pulls a new wire each time a phone is added, and just clips the ends and leaves it in place when a phone is pulled. Eventually, so the stories go, the ceilings collapse under the weight.

Now wouldn't you be embarrassed if this happened to you?

If the PBX is the brains of your telephone system, cabling is its lower digestive tract. Cable management is the most thankless, annoying, tedious aspect of telephone system management.

It is also vital, and if you lose control you'll never get it back.

Why Do You Care: Fundamentally, you need to keep track of cable to know when you have to pull new wire and when you don't. If you know which pairs are in use and which ones aren't, when you add or move users your technicians can get the job done efficiently. If you lose track, they'll have to physically test each wire in the closet to find an unused pair.

You also care because when problems occur, you need to know which cables are involved in order to diagnose and fix them. Again, you can always resort to the, "Hey, Charlie, which pair is moving?" approach. It's faster and easier, and therefore cheaper, if you can just look it up.

If you really lose control, and go to the "clip and pull" philosophy, you'll care because eventually you'll run out of room in your risers (large vertical pipelines that run cables between floors).

The Easiest Solution: If you're contracting out any portion of your phone system management, make sure your vendor takes care of cable management as part of the deal. Review the vendor's cable management process before you sign the contract; make sure you own all rights to the information, and periodically review the cable records to make sure they're up to date and accurate.

A Few Fundamentals: Not everyone knows, or cares, how buildings are cabled. You certainly don't need to become an expert. In any semi-major metropolitan area you'll find dozens of cabling companies vying for your business. They know the building codes and technical specifications for any kind of wiring you'd ever want in your building, and they're engaged in cutthroat competition with each other. Take advantage of the situation, and don't bother taking care of cabling yourself.

Here's what you *do* need to know:

- Analog phones generally use two pairs of wires, called the E&M lines. Legend has it that *E* is for the *e* in "receive"; *M* is for the *m* in "transmit." Actually, they stand for "Ear and Mouth."
- You'll also hear about "Tip and Ring." These are voltage and ground lines, respectively, and refer back to manual switchboards and their quarter-inch diameter plugs. The contact made by the tip of those plugs contained the signal voltage; the ring that made contact with the shank of the plug carried the ground reference.
- Digital phones generally use two pairs, and sometimes three pairs of wire. Unlike analog phones, you'll find no single standard for digital telephone wiring.
- Cable is pulled in bundles of 25, 50, or 100 wire pairs.
- Except for the lines going to individual offices, all wires are terminated (that is, they end) in punch-down blocks. All wires are labeled at both ends, at their points of termination. (If you're not familiar with punch-down blocks, they are simply a means of connecting wires without having to physically twist them together and tape them.)
- The punch-down blocks are located in wiring closets which, in well-designed buildings, are conveniently located to make the job of cabling individual offices, and running cable to your PBX, easy.
- Your building has either a central wiring closet, or its equivalent, located in the switch room. The term means just what it implies—cable begins here, and fans out, closet to closet, in hierarchical fashion, until it reaches each office. There are no voice cables anywhere that don't eventually connect to a termination in the central wiring closet.

  More technically, the group of punch-down blocks in your central wiring closet is called the *Main Distribution Frame.* In a typical building cables will fan out from here to *Intermediate Distribution Frames,* from these to local wiring closets, and from local wiring closets to individual offices.

- Cables run from floor to floor through risers—typically, four inch diameter pipes.
- Cables run from wiring closets to offices in one of the following ways:
  - Hung from a "ladder" (that's what it looks like) attached to the real ceiling and hidden from view by a suspended ceiling. From the ceiling, it reaches the office through a "power pole," which is just a hollow pipe of some sort that reaches from floor to ceiling.
  - Dumped across the top of a suspended ceiling, then through a power pole.
  - Run through a "tray" system in the floor (literally, troughs or pipes embedded in the floor to accommodate cable).
  - Run under a raised floor.
  - Run through "panduit"—a shell you can glue onto a wall so the wiring doesn't look too ugly.

Now you know enough to be dangerous.

Keeping Track on Paper

All you really need is a notebook that contains a line for each pair in each wiring closet. Simple, no? Use ledger pages as shown in Figure 6.

Use a separate group of ledger pages for each wiring closet, and put in a tab separator for each closet. For each pair that originates in the closet, enter the cable number, the closet to which it runs, and the number of the cable to which it is cross-connected. For cables that run to phone jacks, enter the office location instead of a wiring closet, and the telephone extension instead of a cross-connect cable number.

Instead of putting everything in a tabbed notebook, you may choose to keep your documentation in the wiring closets themselves. Two advantages to this: Your technicians can always find the documentation when and where they need it, instead of having to track down the notebook; and, they'll have an easier time keeping documentation up to date, since it's right in front of them when they modify the wiring. The disadvantage: You don't have your information in one place where it's easy to find.

Pick one. Or, you may be ambitious enough to do both, and hope both sets stay in sync with each other. If you do,

Telephone Cable Ledger for Wiring Closet ________

Cable #	To Closet	To Cable #	Cable #	To Closet	To Cable #

Figure 6.
Cable Ledger Page

make sure you perform a documentation review on a quarterly basis, to compare the two and reconcile any differences.

These ledger pages are organized to trace cables down—that is, from the phone system to the phone jack. You can use them to trace cables in the other direction, but it is more difficult. Note that the phone jack cable is recorded in the Extension Ledger as well (to assist in troubleshooting).

To trace a cable from the PBX, simply find the cable number in the System Ledger, then find the matching cable in the Cable Ledger. Since each cable in the Cable Ledger includes the number of the cable to which it attaches, you can go from cable to cable until you find a cable with an extension recorded, indicating the end of the run.

To find unused cables in a wiring closet, simply scan for cables with no entry in the *To* columns.

You'll find instructions for creating a computerized cable database in Appendix A.

PHONE COMPANY CHARGES

Almost as much fun as managing cable is the monthly chore of auditing your bills from the various phone companies with which you do business. If your company does a lot of long-distance calling, or uses WATS service, this little chore can take up a sizable amount of time. With most local phone companies moving to metered service (you pay for each local call, too, instead of a flat monthly fee), a detailed review of your bills is simply impossible, and possibly unnecessary.

Don't review each line of every phone bill. Instead, look at your goals.

Why Do You Care: You may have several reasons:

1. You may need to charge back telephone costs to individual departments (not a bad idea, since it makes each department's management responsible for long-distance costs).
2. You may need to enforce a "no personal use" policy.
3. You may suspect that either an employee or an outsider is using your system for expensive long-distance or international calling.
4. You may suspect the phone company of inaccurate billing.

What You Can Do: If your system has a *Call Detail Reporting* (CDR)—or *Station Message Detail Reporting* (SMDR) port, you can simply attach a printer to it, and let it eat trees. At the end of each month, you can match the CDR log to your phone bills.

Unless your company makes very few calls, this is a horrible approach to call management.

If you need to charge costs back to individual departments, chances are you should have a call management system of some kind. These systems run on personal computers and attach to the CDR port of your PBX (chances are you won't be able to use this sort of arrangement on a key system).

As calls are placed and routed through your system, the call management system logs the CDR data, applies information from rate tables you've previously entered, and calculates the actual charges due from each telephone. Any decent system will provide facilities for rolling up groups of telephones into different organizational structures (departments, divisions, etc.) to let you charge back costs at any organizational level.

A call management system is essential for achieving nearly any of the goals listed above. If you decide you want to do more than simply pay your bills as they come in, get one.

If your only need is to detect unauthorized use, you may get by with simply scanning your bills. If someone made sixty-three calls to Baden-Baden last month, you should be able to pick it up without an automated call management system. (Figuring out who it is may still require a call management system, or at least a CDR log.)

Payoff Idea: Investigate having your local phone company take care of this for you. They'll charge for the service, of course, but they have an advantage you don't—their system knows where calls come from. A legal firm could take advantage of this for billing phone time to customers, in addition to the management aspects of call detail reporting.

Reminder: Change your DISA password frequently. DISA is Direct Inbound System Access and it allows employees to either call your system through an inbound WATS line (from a hotel, for instance), then make a local call; or to make a local call to your system (from a client site) then use one of your outbound WATS lines.

DISA is a popular target for system abuse by outsiders. Once you know the basics, it isn't hard to dial into company systems at random and attempt to gain access—many companies never change the factory default password. Similarly, it isn't hard for a former employee to rationalize continued DISA use (the human mind is wonderfully flexible when it comes to ethics).

Traffic Management

One other use for a call management system is *traffic management.*

The primary economy offered by a PBX or key system over simply bringing a separate business line to each desk is the ability to share outside lines. If you assume not everyone is making an outside call at the same time, you can make do with fewer outside lines than you have telephones.

But how many?

Books—*large* books—have been written on the subject of traffic engineering. You can read one of them, or you can use some simpler rules of thumb.

For a typical small business with twenty-five employees or fewer, a ratio of two telephones for each outside line is about right. (This is called the *trunking ratio.*) If you run a sales office or telemarketing operation, this ratio won't work for you, but it isn't a bad estimate. (For a telemarketing operation use a 1:1 ratio unless you have a fancy schmancy telemarketing system, such as those covered in chapter 9.)

For a larger business (500 phones) an 8:1 or 9:1 ratio tends to work well, again depending on your type of business.

If you're somewhere between these two sizes, figure out where you fall on the trunking ratio graph on the next page.

Why Do You Care: If you have too many outside lines (trunks for a PBX) you're wasting money. If you have too few, calls are blocked and people yell at you.

How Do You Do It: The key buzzword in traffic engineering (aside from trunking ratio) is *level of service,* defined as the percentage of calls that are blocked during peak use peri-

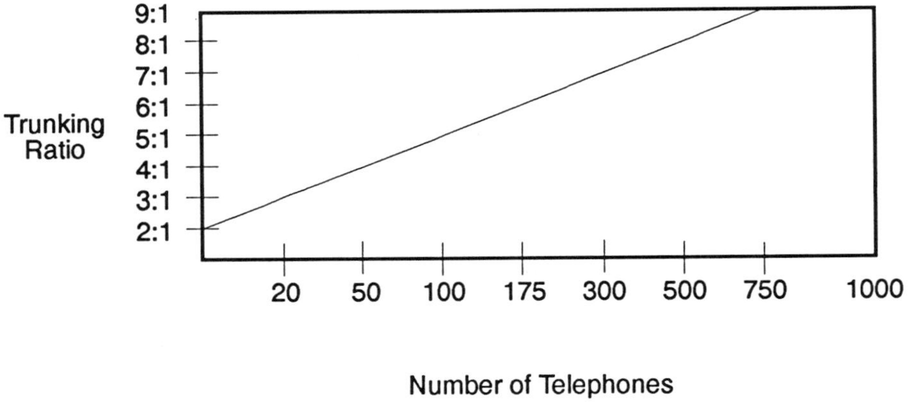

ods. For example, a P1 level of service means that during peak periods, 99 percent of all calls go through.

You need to know this to talk with your vendors. Realistically, you'll never get to be that precise.

You'll know if you don't have enough trunks, because you'll get a lot of gripes. Detecting a trunk surplus is harder.

When you first install your system, work with your PBX and local phone company sales representatives to establish an initial trunking level. They have experts in traffic engineering to help you out—people who like to talk about Erlang Tables and Centrum Call Seconds (CCSs) and things like that.

Remember that the above chart is intended for Kentucky Windage, not precise estimates.

Once your system is up and running, either use your call management system (if you have one) to determine your trunk usage during peak hours, or rely on the following:

How to tell if you don't have enough trunks: Don't worry, you'll know. You'll have *lots* of people telling you their calls aren't getting through.

How to tell if you have too many trunks: If you're nasty, simply cut down on your trunks, one at a time, until complaints are just below the unbearable level.

A better technique is to work with your phone company to find out your trunk usage. They won't be able to give you the kind of detail you could get out of a call management system, but they can probably give you enough information to help you decide if you can safely try a trunking reduction.

Payoff Idea: If you have more than 100 phones, buy a call management system. You can probably get one for less than $10,000, which is what you're probably spending on basic phone service in a year. It will help you adjust your trunking to optimum levels, and will catch any long-distance abusers at the same time.

Work Orders

If you currently manage a phone system, or you've ever been responsible for an operations type of environment, you're probably familiar with some form of work order process. If you haven't had to manage one in the past, here are some guidelines.

1. If your only goals are operational, use a paper system. Computerization adds a lot of ability for management reporting—which may or may not be important to you—but doesn't add much value to the process of controlling the work itself.

2. If you're doing all the work yourself, or you're contracting out all work, formal work orders are overkill. Use simple To Do lists, or Post-It® notes.

3. Decide whether your work order form is for you and your staff only, or for your end-users to fill out. If it's for your end-users, make it as simple and generic as possible. If your work order form is for internal use, emphasize efficiency over simplicity since it will be used by you and your trained staff only.

4. If you've decided on having your end-users fill out the form, consider having them submit short memos instead, since forms are intrinsically intimidating. Forms do give you the advantage of a standard format for all requests, so you have a decision to make. That's why they pay you the big bucks in management.

5. Someone should review all open work orders based on your standard resolution time. In other words, if you plan on completing work orders within twenty-four hours, review them daily.

Keeping Track on Paper

To set up a work order system on paper, first (of course) design the form. Figures 7 and 8 show typical work orders for customer and internal use, respectively. Tailor them so they make the most sense for your organization and telephone system.

Regardless of the form you use, have the printer give you prenumbered, carbonless forms prepunched for standard notebooks. The second sheet should be stiff enough to keep writing from penetrating to the next form.

You want two copies of each work order, with the same number preprinted on each. One goes in the in-basket of the technician who will complete the work. The other goes in a notebook you will set up to hold open work orders.

When work is completed, technicians will initial their copies of the work orders, write in the date and time of completion, and place them in your in-basket. You take that copy, staple it to the original, and transfer them to the completed work order notebook.

Very simple. Low overhead. Effective.

Obviously, you need to computerize instead.

To be fair, you may have requirements for management reporting that make a computerized system essential. We'll try to design it to be no more cumbersome than the paper-based system. If you're interested—and stout-hearted—refer to Appendix A for details.

One More Thing

If you've waded through all of the above, you already know most of what it takes to select, install, and manage a telephone system.

You've been using telephones all your life. Once you know the basics, taking care of telephone systems requires little more than common sense, multiplied with a lot of patience.

And training.

Because every system is different, this book glosses over the details of telephone system administration. This doesn't mean there is no training required. Quite the opposite. Unless you can contract out your system management, you'll find you

Telecommunications Work Order Date: ___/___/___ # ________

User Name: ______________ ______________ Department: __________

Last Name First Name

Location: ______________________________ Contact: __________

______________________________ Phone: __________

Phone: ______________

Work Needed:

Date Needed: __/__/__

INTERNAL USE ONLY

Assigned To: ________________

Comments:

Parts Used	
Part	Qty

Completed (Date/Time): Initials:

Figure 7.
Customer Work Order Form

Telecommunications Work Order Date: ___/___/___ # ________

User Name: ______________ ________________ Department: __________

Last Name First Name

Location: ____________________________ Contact: __________

____________________________ Phone: __________

Phone: ______________

[ ] Repair [ ] Remove
[ ] Chg Class to ______________________
[ ] Chg Group to ______________________
[ ] Chg Model to ______________________
[ ] Other
(attach details)

[ ] New Ext: ____________
[ ] Move — New Location:
___ ________________________
___ ________________________

Date Needed: __/__/__

INTERNAL USE ONLY

Assigned To: _______________

Comments:

Parts Used

Part	Qty

Completed (Date/Time): Initials:

Figure 8.
Internal Work Order Form

never know quite enough about administering your system to fix the latest problem.

Finally. . .Every so often, take a moment to reflect on just how astounding this all is. We live in a science fiction age, and it is nowhere more apparent than in how commonplace telephones have become. In seconds, we can talk with people anywhere in the world, using copper wires, glass fibers, or satellites, and it isn't often we take an extra moment to think about it all.

So as you carefully enter the route of a new cable in your logbook, take a second to marvel at it. You're managing a miracle.

CHAPTER FIVE

Understanding Telephone Services

YOU CAN PICK UP YOUR PHONE and call anywhere in the world, a fact that never ceases to amaze. The bill you pay for this service also doesn't cease to amaze—it's really quite low, considering what you're getting.

That doesn't mean it wouldn't be even better if it were lower.

You buy telephone service from one local carrier and at least one long-distance carrier. You rarely have a choice regarding whom you buy local service from (although that may change in the near future). You still have a lot of decisions to make regarding *what* you buy.

Long-distance is a competitive arena, so you have to decide who you want to do business with, then decide what services you want.

Multiple choice quiz time:

Determining the elements of local telephone service is:

A) Cut-and-dried.

B) Straightforward, but with some areas for creativity.

C) Complicated but understandable. Once you understand the rules, the decisions are unambiguous.

D) A black art with no apparent logic or basis.

E) All of the above.

The answer, of course, is E) All of the above. If your business is relatively small and you use a typical key system then your decisions are mostly cut-and-dried—you'll lease enough 1FB lines (standard analog lines) from your local phone carrier to make sure you don't see all the lines lit at once.

If you are a larger business, with a PBX servicing a thousand phones, then you have more decisions to make.

For that matter, you may have decided to use Centrex or Centron service instead of buying your own telephone system. If so, your decisions will be easier, since your telephone company will make them for you.

Don't downplay the value of being taken care of. Many companies that could use Centron service to great advantage purchase their own PBX out of a false sense of economy. Owning your own system carries several hidden costs, one of which is administrative effort.

Payoff Idea: Consider using Centron service for disaster recovery. With proper planning, (and additional costs) you can avoid losing more than a day of service even if you lose your PBX, or even if your whole facility is lost to a fire.

You can buy big, expensive books on the subject of disaster recovery planning, so we won't go into any detail on the subject. Work with your local phone company representative so you at least have an idea of how you'll get back in business if you lose your phone system or your whole facility.

Types of Local Telephone Service

Let's begin with a primer on the types of local phone service you can buy. First, a question: What is the difference between a telephone line and a trunk?

Answer: A trunk costs more.

That isn't really fair. As far as your local carrier is concerned, a trunk comes out of a completely different part of the central office switch. Electrically, though, there isn't much difference.

Another question: What is the difference between a business line and a household line?

Answer: A business line costs more.

Don't make a big fuss about this, though. American business benefits greatly from having telephones in nearly every household in the United States.

1FB service is the phone company's name for a standard analog business line—the kind of line that runs into a phone jack on the wall, like you have at home.

Key systems use 1FB lines, and aggregate them into multibutton telephones so each telephone can use any of several different incoming lines.

If you're installing a key system, make sure your 1FB lines are configured in a *hunt group.* In other words, when one line is in use calls are automatically routed to the next line.

How many lines should you order? Chapter 3 provided some guidelines, which we'll repeat here. For a typical key system, expect to need about one line for every two employee telephones, but be reasonable about following this guideline.

Example: You have two employees, so you decide to follow the above advice and order just one line. Right. Don't be a cheapskate—order three lines (one for yourself, too. Don't buy a key system, though—Centrex service will be just fine for your business.)

Example: You run a sales office, city desk, or telemarketing operation with twenty employees. You follow the 2:1 guideline and order ten lines.

No, you don't; you order twenty lines. Be reasonable. Use the 2:1 guideline (3:1 for a larger operation) for a typical office, not as a one-size-fits-all solution.

Payoff Idea: Make sure you order 1FB lines with separate numbers assigned to each (they don't automatically come this way). You don't really need this for daily use, since you'll publicize only the first number in the hunt group, but it can help a great deal if you or your vendor has to track down a problem with your phones.

Another Payoff Idea: You can easily give yourself a private line (or your boss, if you're only a medium-sized cheese in your organization). Simply ask for all your numbers *except one* in sequence, and leave that one out of the hunt group. For example, you may get the numbers 555-4010, 4011, 4012, 4013, and 7989. The big cheese gets 555-7989 as a private line, even though it runs through your key system and is usable in a pinch for outgoing calls. You can even leave it off most of the office phones.

Trunk is a generic term, used to describe phone lines that attach to PBXs. If this book is your first exposure to modern telephony you may think that a trunk is a thick bundle of cable.

It sure sounds like it, but a trunk is actually just a single voice circuit—it can handle only one conversation at a time.

Before we go into any specifics at all, you need to make a basic business decision. Do you want DID (Direct Inward Dial) service, or do you want an operator (or automated attendant) to route calls?

DID means that you can give every phone a different outside calling number so that callers can reach employees by dialing their numbers. The alternative is to use a central number, and give employees their own extensions.

So how are DID trunks different from simple incoming lines or non-DID trunks? Each incoming line (like those attached to key systems) has its own number. A key system allows you to pick up that line from many different phones.

Regular inbound trunks have no assigned number. Any call for your central phone number can come in on any trunk.

With *DID trunks,* calls to a specific number may come in on *any* DID trunk. Embedded in the signal that comes from the central office is the called number, allowing your PBX to route the call.

There are no guidelines for deciding on DID service instead of a central number with extensions. Keep in mind that even with DID service and an automated attendant, you'll still need an operator to handle callers that don't have an employee's DID number.

Also keep in mind that DID service costs more. On the other hand, it takes a sizable load off your operators and shortens calls (fewer callers are put on hold) so you'll need fewer trunks. It also will make you more popular with your customers—instead of an operator putting them on hold, they can call straight in.

Opinion: Use DID service, at least for your professional staff.

Analog trunks are electrically no different from 1FB trunks, but they are tariffed (read priced) differently, and emerge from a different section of the central office switch. If you want detailed information on the why's and wherefore's, talk to your phone company representative.

Trunks attach to PBXs; 1FBs attach to key systems.

Analog trunks come in three flavors: inbound (you can use them to receive calls, but not to initiate them); outbound (you can call but nobody can call you); or two-way (you can use them for both incoming and outgoing calls).

One restriction you need to know about: You can only get DID service on inbound trunks. It isn't availabe on two-way trunks, and doesn't mean anything at all in the context of outbound trunks. (You figured that one out by yourself, didn't you. See how easy this business is?)

Digital Switched Service (DSS, although different phone companies use different names) arrives at your PBX as a T1 circuit. T1 service is a digital transmission standard. All you need to know is that it breaks down into twenty-four digital trunks. (We'll revisit T1 many times in this book. Each time you'll learn something new that you never cared about before.)

You don't have to lease all twenty-four trunks. In other words, the the phone company attaches capacity in twenty-four-trunk increments, but you buy it one trunk at a time.

Unlike analog trunks, digital trunks are all two-way trunks, and don't have any restriction regarding DID service. Digital service has several other important characteristics:

1. Its sound quality is vastly superior to analog service—essentially noise free. (You can still get noisy calls, since the phone you're calling may have analog service).
2. It is "faster." (No, your voice won't travel more quickly.) With analog service, after you hang up your phone there is a lag before the trunk you were using is ready again. Digital trunks are ready faster, which in theory means you need fewer of them.
3. It costs more. Ask your local phone company for details, since rates are tariffed, and each local Public Utilities Commission negotiates its own rates with the phone companies.

ISDN service (ISDN stands for "Integrated Services Digital Network," although some claim the translation should be

"It Still Dials Numbers") comes in two flavors: *BRI* (Basic Rate Interface) and *PRI* (Primary Rate Interface).

BRI is also known as 2B+D service. In other words, it provides 2 *B* channels and 1 *D* channel. Is that more clear?

How about this: The D channel is the signalling channel, used for internal control information and also for sending data. B channels are short for bearer channels—they carry the content.

What You Really Need to Know About BRI service: This is how you get Centron service with ISDN features; and with BRI each phone can handle two simultaneous calls and independently handle data transmission. (If you care about this, read about it in the companion volume to this book, *Computer Systems for Every Business.* It gets complicated very quickly.)

PRI is also known as 23B+D service. Now that you know what B and D stand for (your life will never be the same again. . .) you can guess that one PRI can deliver twenty-three voice conversations plus a bunch of data transmission.

If you want ISDN on your PBX, you usually buy PRI, although you can also bring BRI service into a PBX.

Can you buy PRI service on a trunk-by-trunk basis? Ask your phone company. This is a rapidly-changing area, and varies on a regional basis. You probably can't, since the phone company can sell you BRI service to accomplish the same purpose.

Proponents of ISDN can show you a list of impressive features. The important ones are *automatic number identification* (ANI) and *simultaneous voice and data.* And ANI isn't unique to ISDN—you can receive it through "in-band signalling" on standard analog trunks.

In the future, voice/data integration will probably cause ISDN to succeed. (More cynically—ISDN will succeed because the phone companies can't afford for it to fail. *They need* a standard for digital interconnection, and want you to pay for it.)

Payoff Idea: With brand-new technology, it usually pays to wait for the future. The future is cheaper. In other words, buy ISDN now if you need ISDN now, or if you expect to next year. Don't buy it because you know you'll buy it sometime, unless your phone company makes it financially attractive for you to do so.

Stay on the leading edge, not on the bleeding edge. Right now, ISDN is closer to being the latter than the former. Keep an eye on it, though.

Trunk Engineering

Terms like *trunk engineering* sound precise and abstruse, don't they? In the real world, you fiddle around until it comes out right.

What You Need to Do: Decide how many inbound, two-way, and outbound trunks you need. You want a mix because most businesses don't want a peak period of incoming calls to block their ability to call out, nor do they want customers calling in to receive busy signals.

Basic Rule of Thumb: Use a 1:3:1 ratio of inbound, two-way, and outbound trunks. If you figure you need a total of 100 trunks (servicing somewhere in the neighborhood of 900 telephones) you should have about 20 inbound trunks, 60 two-way trunks, and 20 outbound trunks. You call these things *trunk groups* (now you know how jargon gets started).

If you use DID service, the two-way trunks won't do much for you, since you can't put DID service on them. With DID service, a 1:1 ratio of DID to outbound trunks is your best bet, unless you know in advance that the ratio of inbound to outbound calls is skewed.

If you buy digital service, have your central office provide (continuing the above example) eighty two-way trunks and twenty outbound trunks. In your PBX, define twenty of your two-way trunks as inbound—in other words, the PBX won't direct any outbound calls their way. This trick makes it easier for you to fine-tune your system without ordering changes from the phone company.

Power Failure Phones and Trunks

If phone service is critical enough to your operation, you may want to consider having some *power failure phones* scattered throughout your offices. Power failure phones are special phones that can still be used if your PBX stops working—in case you experience, well, a power failure, for example.

Suggestion: Use red telephones for your power failure phones, and don't have any other red phones to confuse them with.

Typically, power failure phones connect to analog ground-start lines which double as outbound trunks. Make sure these come from a different central office than your main group of trunks, to protect you from a central office failure (central offices do fail occasionally).

Power failure trunks will form part of your pool of outbound trunks, but if you have digital service for the rest of your trunks, make sure your PBX addresses the power failure trunks *after* the other outbound trunks, but before the two-way trunks. Since analog trunks provide poorer sound quality, use them last.

How to Use Trunk Engineering Principles with Key Systems

Since the basic principles of trunk engineering depend on dividing trunks into inbound, two-way, and outbound groups, you need to mimic this process with your 1FB lines. It isn't hard (you may have it figured out already) but just to make sure. . .

Emulate *inbound* trunks through office procedures—ask your employees to avoid several lines when placing calls. If you have twenty lines, starting with 555-4400, ask them to avoid using 4400 through 4404.

You don't have to emulate *two-way* trunks: 1FB lines are all two-way lines.

For *outbound* trunks, leave a few numbers out of your central office hunt group—if you have twenty lines, leave about four lines out. You'll still be able to receive sixteen simultaneous incoming calls, but you'll have four lines reserved for outbound calls.

Trunking Analysis

So far, you have rules of thumb to work with. But how do you fine-tune the system so you have the best service at the lowest cost?

After you've used your system for several months, ask your phone company to run a traffic study. Traffic studies aren't free, but they are reasonable (ask your phone company representative for a quote). For each hour of the day, a traffic study will tell you the number of calls coming in and out of your phone system, the peak and average number of trunks in use, and the number of times callers experienced busy signals.

Pretty basic stuff. Everything you really need to know.

What do you do with this information? You adjust your trunking, of course. If all trunks are in use during peak periods, you should probably add a few trunks. If not all trunks are in use but callers are experiencing busy signals, switch a few outbound trunks to two-way service. If you have no busies and never have all trunks in use, you can reduce your trunking.

Flat-Rate or Measured Service

Some areas still have flat-rate service available, but it is rapidly going the way of the dodo bird and the assumable mortgage. Flat-rate service is what we all grew up with: You pay a monthly fee and get all your local calling at no additional charge.

Measured service, also called "metered service," means you pay for the service you use. Metered service comes in several flavors, including one with a monthly connection fee that covers the first *n* calls; you pay for any calls above that number.

The details of what your local phone company can offer depend on your local Public Utilities Commission.

Which Should You Use? You may have no choice. Depending on where you do business, flat-rate service may not be available. Even if it is, it may not be the most economical approach for your business (although it *is* the most predictable).

Pick some kind of service and live with it for a few months. Then ask your sales representative to do an analysis for you, to show you which form of service would minimize your costs. Based on your call volume, you will be able to see which package makes the most sense.

There is no one-size-fits-all solution. The best advice: Don't assume.

Special Features

Your local phone company has a thick book of special features they can provide you, and every phone company's list is different. Here is a smattering of the more useful ones:

1. **Custom Numbers** are familiar, but this doesn't mean you shouldn't have one. You can either use the last four digits ("For the best investment advice, call 555-BOND.") or all seven digits ("For the best dining experience in town, call EAT-GOOD.").

   Remember, the first three digits (the exchange) are determined by the central office switch providing your service. If you need a 328-number ("EAT") but 328- doesn't service your location, you have two choices. Either you can change your strategy to go along with an available exchange, or you can spend more for a remote call forward or foreign exchange (see below) and get the number you want (unless someone else is using it already).

2. **Remote Call Forward** simply means the phone company redirects any calls from one number to another number. This service has two main uses: Number changes and custom numbers.

   If your business has to change phone numbers, you may want to keep your old number in service for awhile. Doing so gives your customers time to adapt, and preserves the current phone directory's integrity. **Suggestion:** Use remote call forward until new directories are published. A remote call forward switches customers who call your old number to your new number. Since there is a charge for this service, keep it as long as you like.

3. **Foreign Exchange** is not having a Moldavian teenager stay with your family for the summer, although that can be a valuable experience too. In this context a foreign exchange is, literally, phone service from an exchange foreign to your location. In other words, you can run a trunk directly from a central office in another city into your PBX.

   But why? You may want to have different sales groups handle calls from different suburbs. If suburban calls come in on different trunks, you can have your PBX direct them to a separate group of phones.

   Or, you may want a custom number (see above).

   Foreign exchange service also gives your customer a local call, which otherwise might be long-distance. (Remember, your local carrier also charges long-distance fees for calls outside your local calling area.)

4. **Call Detail Reporting** (CDR, also known as station message detail reporting or SMDR) was discussed in detail in the previous chapter. Remember that your phone company may be able to do this for you. Go over the details with your representative, especially to discuss how the company handles such details as tracking long-distance service. (**A Guarantee:** If your long-distance carrier provides trunks directly to your PBX, your local phone company won't be able to report this traffic.)
5. **976 service** is available to you also—local callers pay to talk to you about something, and you split the take with the phone company. Dial-a-porn services have tainted 976 numbers, and 900 long-distance gives you the same feature nationwide.

   Opinion: Don't use 976 unless your business does dial-a-porn (we get the most diverse readers...) and be careful about 900. See page 83 for more.
6. **Automatic Number Identification** (ANI) is becoming available in some metropolitan areas. ANI will be of greatest value in conjunction with interactive voice response technology, which you'll find explained in chapter 10.

Fiber Optic Communications

Anyone with a television has seen a technician listen to the sound of a pin dropping on a glass desktop, demonstrating the incredible quality of fiber optic communications systems. Impressive, isn't it?

It's also baloney.

Yes, fiber optic communications are effectively noise-free and clean. On the other hand, *so is any other form of digital transmission.* If you have an end-to-end digital signal path, you too will be able to hear a pin drop.

Payoff Idea: Don't pay more for fiber. Buy digital service, unless fiber costs the same.

A Few Final Words on Local Phone Service

Local phone companies constantly offer new services, new flavors of old services, or new ways of packaging and pricing groups of services. As a result, the most cost-effective way of obtaining service today may not be the most cost-effective way tomorrow.

Make it a habit to sit down with your phone company representatives every six months or so. Use them as free consultants. If you have a harebrained notion of something you might do with your phone service, don't be shy—ask. You never know what may be available.

Long-Distance Telephone Service

There may be no area of telephony harder to keep track of than long-distance service. Don't try, unless you enjoy this sort of thing as a spectator sport.

The rules keep changing, since regulation of the phone industry has turned into full-time employment for federal judge Harold Green. Think of him as the federal equivalent of your local Public Utilities Commission and you won't be too far off base.

Your requirements for long-distance service are very simple: The highest-quality service at the lowest cost. That's all you care about. Now, let's decide what we mean by quality.

You care about three things: Sound quality, billing accuracy, and problem resolution. Any company that gets these right doesn't have to worry about anything else.

Sound Quality

Three components contribute to sound quality: Your local access to the long-distance network, the network itself, and the other party's local access.

Local Access: If you don't do anything special, you connect to long distance through your local phone company. When you dial (for example) 1-312, your phone system connects you to your central office, which then connects to your default long-distance carrier's network, which establishes a connection to a central office in Chicago when you finish dialing the number.

You could use the finest long-distance network in the world (you'll probably use one of the three best anyway) and still experience a lot of noise. Why? You may have local analog service and get a noisy trunk, or the person you're calling may have local analog service and may get a noisy trunk. Or, depending where you are, your local phone company may have an analog connection to the long-distance carrier.

Analog service means uncertain sound quality.

Payoff Idea: If your company has a PBX and does a lot of long-distance calling, your long-distance carrier(s) usually can bring a trunk directly to your facility, bypassing your local carrier. Since you eliminate a variable, you improve quality. Depending on how your local carrier charges, you may save on local access charges as well. Also, connections will be faster. Work with your long-distance carrier to determine your trunking requirements.

As with local service, you can connect to your long-distance service using analog, T1-digital, or ISDN service. **Opinion**: Unless the savings for analog service are compelling, go with T1 service. The improvement in voice quality will be worth it. As of this writing, you should probably hold off on ISDN. You can always convert later, when the technology has matured and costs have come down.

The great thing about direct connection to your long-distance service is that you lose nothing. If you don't have enough trunks you can still get long-distance service through your local carrier. Figure 9 shows how it works.

Question: Do you worry about a long-distance provider's financial stability?

Answer: Only if you hold stock in the company.

With your phone system you care a great deal, because you need repair parts and trained technicians. If your long-distance provider goes toes-up, however, you can switch to another without tremendous disruption.

(Not, however, with *no* disruption. You'd stand a good chance of losing the 800 numbers your customers have grown accustomed to.)

Before going any further, let's take a quick look at long-distance service, just in case. As usual, prepare to see things you already know about. Also as usual, expect a few new nuggets scattered about the old gravel.

1. **Standard Service:** You dial the phone, you reach the number, you pay the bill. Someone dials *your* number, reaches your

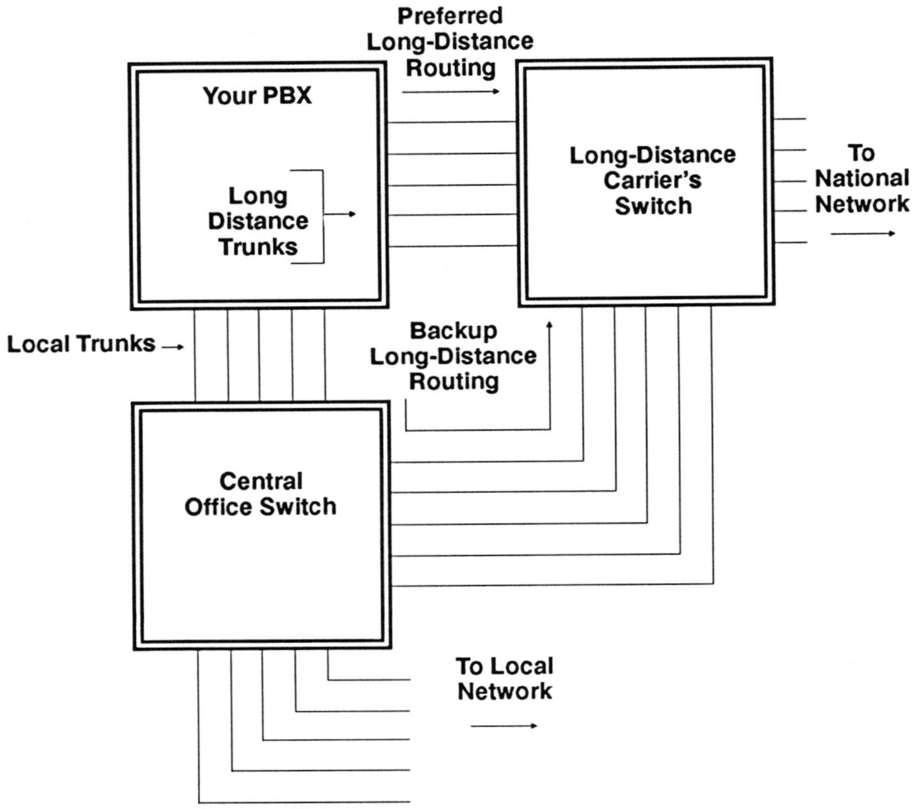

Figure 9

phone, pays the bill. Nothing fancy, except the technology behind the scences, which is awesome. Someday, if you have the time, read about it.

When you make decisions regarding long-distance, keep in mind that different providers may provide lower costs to different areas. If your system comes with automatic route selection (see page 36) it can pick the lowest cost provider for each outgoing call.

2. **Inbound WATS:** WATS is another acronym whose origin (Wide Area Transport Service) nobody cares about. Originally, AT&T's idea was to provide a flat charge for unlimited service. That changed to the concept of a uniform charge, regardless

of distance. Now we have distance-sensitive WATS, an oxymoron of sorts.

Inbound WATS is 800 service. The key to inbound WATS is that you pay for the call, even though you didn't originate it. It's a great thing to provide your customers (if you are a consumer-oriented business, you really have no choice).

Payoff Idea: Employees like to be big-shots, and may give out your 800 numbers to people other than customers—vendors, for example. Rather than fight with your employees about this, have your 800 numbers route directly to the sales department rather than your general number, if at all possible.

3. **Outbound WATS:** See above. By now, outbound WATS is just another name for discounted long-distance service. Not that there is anything wrong with reducing your costs. Don't expect a flat-fee, or a one-price-fits-all cost structure though. You may still find one out there, but it isn't likely.

4. **900 Service:** With a normal long-distance call, your customer pays the phone company to talk to you. With 800 service, *you* pay the phone company so your customer can talk to you. 900 service means your customer pays through the nose to talk to you, and you split the take with the phone company.

   900 service is becoming popular for a variety of services, from special call-in promotions to service departments. You may find some uses for it yourself.

   Payoff Idea: Don't be greedy—search your soul to make sure there is enough value *in your customer's mind* to warrant the cost.

Opinion: One of the worst ideas in recent history is the use of 900 service to defray the cost of customer service departments, a practice becoming increasingly common among companies that sell technical products such as software. Before you decide to go this route, think from your customers' perspective: They're calling because your product doesn't work, and you want to charge them to fix it.

Even Worse Idea: Having your 900 number terminate in a automated call distributor (ACD, discussed in chapter 8). Whoever had this bright idea didn't stop to consider the psychological effect. Charging your customers for the time you put them on hold may be one of the most effective ways yet invented for donating customers to your competition.

LONG-DISTANCE PROVIDERS

You can buy long-distance service from four types of companies. First, your *local phone company* provides something called *Intra-LATA Long-Distance*. LATA means something like "Local Area Telephone Access." Intra-LATA long-distance means you're charged for long-distance but it's all within your local phone company's network.

You can buy long-distance service from one or more *long-distance carriers*—the big three are AT&T, US Sprint, and MCI. These companies have their own nationwide networks, as you undoubtedly know from their extensive advertising campaigns.

You can also buy long-distance service from two other types of companies: Aggregators and resellers. Neither aggregators nor resellers have any networks of their own. Both provide you with service from one or more of the big three (and maybe some fringe players as well).

Aggregators, as the name implies, connect your phone system to their local facility. They buy service from multiple long-distance vendors at volume discounts and switch your calls to the lowest-cost carrier for each of your calls. The long-distance carriers bill them, they bill you.

Resellers resemble aggregators, but without any facility of their own. They simply arrange for volume discounts from one or more of the big three, then pass the discount to you. The long-distance carriers bill you directly, as does the reseller.

Selecting a vendor isn't very complicated. Go with whoever gives you the lowest cost, and whose bills you'll be able to understand.

Determining who will give you the lowest cost can be an interesting exercise in tedium, but it isn't especially difficult, at least in principle. Long-distance services divide your calling into "zones," which are concentric bands on the map. They bill calls to each zone at a different rate. Simply go through a few months of your current long-distance bills and figure your number of calling minutes to each zone, then multiply by the rate for that zone.

Payoff Idea: Don't assume you have to buy all your long-distance service from the same vendor. You may find that different vendors provide better prices in different zones. Feel free to mix and match. The automatic route selection feature of your PBX will automatically select the right service for each call. **Caveat**: You may find yourself losing more in vanished volume discounts than you gain in reduced rates. Be careful how you perform your comparisons.

Another Payoff Idea: Don't assume intra-LATA long-distance is less expensive than inter-LATA long-distance. Cases where calls to the next county cost more than calls to the far coast are legendary in this business. Often you can save money by using a long-distance carrier to handle your intra-LATA long-distance.

Yet Another Payoff Idea: If you have different sales groups for different geographical regions, most PBXs have the ability to route inbound WATS calls directly to the proper group. You win twice: One less transfer for your customer means better service for them, and shorter calls for you (remember, you're paying for the call).

Fiber Optic Communication

For you, fiber optics are no more important for long-distance service than for local service. Fiber optics are incredibly important for the carrier. For you, as long as your carrier has an all digital network, voice quality is guaranteed.

Long-distance is, if anything, less complicated than local service, except you pay more. The key rule is to avoid assumptions regarding price. For companies that do a lot of long-distance calling, the difference between using the default carrier and optimizing service can be as much as 25 percent, and 10 percent reductions are common.

SECTION II

THE FANCY STUFF: ADJUNCT PROCESSORS

So far, we've dealt with familiar stuff, and worked from a theme of, "You're going to have telephones. How can you get the most service at the lowest cost?"

Now we get to the fun part.

Consultants will use phrases like, "You should plan on leveraging your telephone system into a strategic, competitive advantage. . . " which is why they're consultants and not ordinary members of society. Still, there is an element of truth in this overblown phrasing.

You can add a variety of *adjunct processors* to your basic telephone system that will let you deal faster and more effectively with more callers, provide your customers with better service, and save yourself money.

Who could ask for more out of life than that?

Don't think you have to throw out your current phone system, either. Manufacturers of adjunct processors aren't foolish enough to avoid such a huge potential market. Unless your phone system uses tin-can-and-string technology, chances are you can enhance it with the systems described in this chapter. You may have to expand your system, since some adjunct processors increase the number of calls you process, but that's good—more calls generally mean more business. The next chapters deal with:

1. **Autoattendant:** A disembodied voice that answers all calls to your central switchboard number and asks for the appropriate extension. Autoattendants may also present *voice menus* ("If you want to speak to a doctor, press 1. If you would like to speak to a nurse, press 2. If you need a mortician, press 3. . . "). They may do a combination of the two ("If you know the extension, please enter it now. Otherwise, if you want to speak to. . . "). Autoattendant is short for "Automated Attendant" and if you remember that the long form of "operator" is "console attendant," autoattendant becomes self-explanatory.
2. **Voice Mail:** Also known as voice messaging by those who need to keep the price high, voice mail is the fancy equivalent to having answering machines all over the place. Don't be too cynical about voice mail—it can help you perform many useful chores that answering machines can't, so long as your employees are willing to learn how.
3. **Universal Call Distributors and Automated Call Distributors (UCDs and ACDs):** These are the potentially annoying devices

that say, "All of our agents are busy right now. When they are done with the dozen calls stacked up before you, they'll get to you, too. While you're waiting, here is some irritating elevator music to further frustrate you." But don't blame the ACDs. Used properly, they lead to great customer satisfaction. Customers can get through to you during peak call periods, and know you're treating them fairly. Just don't make them hold the phone too long.

4. **Power Dialers/Telemarketing Systems:** Also known as outbound ACDs. From simple PC-based programs to powerful integrated telemarketing management systems, these greatly increase the success of any telemarketing group. Power dialers simply dial phone numbers from a list automatically, saving the time and effort of reviewing a list and punching in the numbers by hand. Telemarketing systems predict when the next agent will be available and dial accordingly, accept sales orders, compute commissions, and otherwise act as complete management systems for telemarketing operations.

5. **Voice Response Units (VRUs):** If you've ever heard people talk about integrating voice and data, well, this is what they're talking about. VRUs ask questions which callers can answer by pressing the buttons on their touchtone phones. The VRU can translate the button press into the equivalent number, feed that number to a computer system, and translate the computer's response back to speech—hence the name voice response. VRUs are blockbuster technology for some businesses—they are becoming the telephone equivalent of Automated Teller Machines.

Of these five technologies, only the last one is remotely near the bleeding edge. You don't have to worry about being a pioneer. This is well-explored territory (although not so well that you'll be the last kid on your block to have one).

All these technologies have something very important in common—you can probably offset their cost with hard-dollar reductions in operating costs, and experience the strategic benefits they provide for free. Put another way, if these fit your business you have little risk, except the risk of doing it wrong. (Poor implementation can ruin any idea, no matter how good.)

CHAPTER SIX

Automated Attendants

IF YOU'VE NEVER INTERACTED with an autoattendant, you've probably been living on Mars for the last five years. They aren't difficult to understand, install, or live with. In fact, there is so little to them that you probably won't buy an autoattendant at all.

Instead, you will probably find yourself buying a voice mail system or ACD that provides an autoattendant as part of the package. This strategy has two advantages. First, your autoattendant is now free (not a bad price if you can get it); and second, you have one less box to learn how to run. (No, it isn't really free. You'll need a slightly bigger voice mail system or ACD than you would if you weren't using it for an autoattendant as well.)

Autoattendants aren't especially complicated. So far as your phone system is concerned, the autoattendant is simply a bunch of phones in a hunt or call-pickup group. It can answer calls, interact with callers, and initiate call transfers.

Internally, autoattendants can store one or more recorded messages, detect touchtone signals, and translate touchtone signals using internal lookup tables. In other words, the autoattendant can present a menu of choices (to which the caller responds by pressing 1, 2, 3, or whatever), translate the

Business Summary

Useful Life:	5 years
Cash Cost:	\$0–\$40 per phone system user
Amortized Cost:	\$0–\$10 per phone system user per year
Dedicated Staff:	Optional and fractional
Justification:	Offloads routine calls from operators and reduces average caller time on hold.

Cost Analysis (estimated)

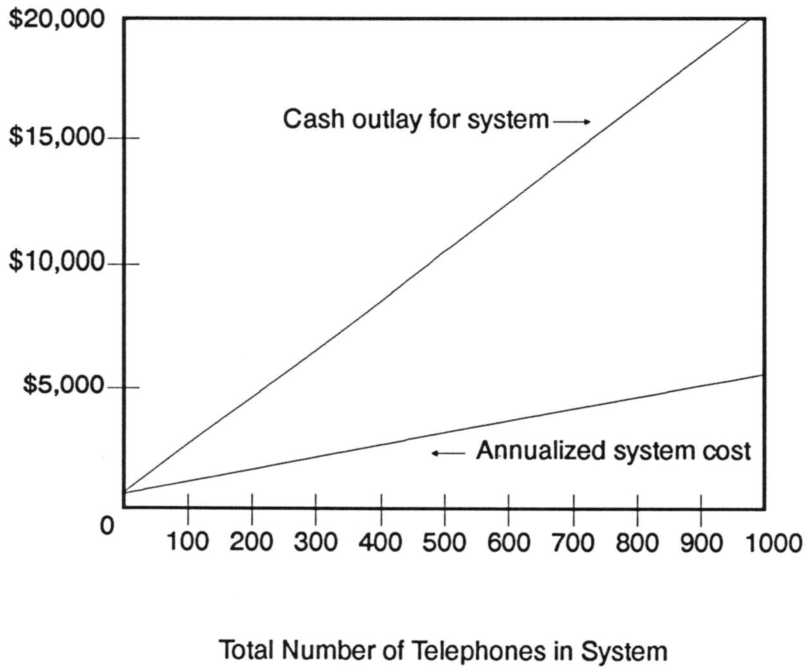

response to a specific extension, and signal the phone system to transfer the call to that extension.

Very Important Point: DID service doesn't eliminate the need for an automated attendant. Even with DID service, you will get a lot of callers who don't know who they should be calling.

YOUR FIRST DECISION

You can buy standalone autoattendants, or you can obtain your autoattendant as a feature of an ACD, VRU, or possibly a voice messaging system. If you need an automated attendant, you first need to decide whether it should be a separate system or not.

Easy Answer: If you can get it for free (as part of another system you're also planning to install), do it that way. You save money and have fewer systems to administer and maintain.

Slightly Harder Answer: Envision how you want your phones answered. A little daydreaming may inform you of a hidden desire for an ACD, or even a VRU, rather than a simple automated attendant.

Use whatever method you used to select your PBX for the rest of your voice systems, including this one (unless the results dissatisfied you). If you are large enough to need an autoattendant you will probably use a formal RFP, but as with the PBX it is your choice.

The Easy Way

You have a phone system of some kind. Ask the company that sold the system to you what autoattendant system they recommend, buy it, and adapt to its features. This strategy has two big advantages: You don't have yet another vendor to deal with; and you don't have to worry about two vendors pointing fingers at each other when something isn't working.

The disadvantages? You may not get exactly what you need (although with a standalone autoattendant there isn't much to leave out); you probably won't get the best price; and you may not get the best system, either.

If you're happy with the service your phone system vendor provides and you want a standalone autoattendant, buy whatever your vendor recommends. Otherwise, do it the hard way—analyze your requirements and research the market.

Your business requirements for an automated attendant are probably uncomplicated. (If you are buying it as part of another system your overall system requirements may be complicated, but the autoattendant part won't be.) Even so, you need to enumerate them for yourself, and for your potential suppliers.

Many of your autoattendant requirements are duplicates of PBX requirements, so don't be surprised when they seem familiar. They may seem familiar for another reason:

You've undoubtedly interacted with these critters in the past, so you have a reasonable idea of what they do.

Requirements

I. Technical Specifications
   A. *Features:* For standalone autoattendants the list isn't long, but there is a list:
      1. *Length of Message:* Are there restrictions on how long a greeting the system will accept? (Of course there are. What are they, and can you live with them?)
      2. *Number of Messages:* How many different greetings can the system store?
      3. *Time-of-day Message Activation:* Can the system automatically switch between different greetings at different times of day? Example: One message for business hours, the other for after hours.
      4. *Number of Autoattendants:* Can the system provide more than one automated attendant, to handle different incoming lines? A business that publishes separate phone numbers for sales, customer service, and general business might want three autoattendants, without having to buy three systems.

         Also, ask about any constraints jointly imposed by limitations in message length, number of messages, and number of autoattendants.

         Don't Say I Didn't Warn You Department: Once you install an automated attendant, your employees are likely to find several additional uses for it beyond the basic need to answer incoming calls. Buy a system with plenty of spare capacity, or at least a lot of expandability.
      5. *Menu Processing:* Does the system let you present a menu of choices, selected by touchtone? Find out how many layers it will allow (having one menu selection invoke a second menu). Also find out how many different menus the system can support concurrently—as with basic greetings, you may want to present different menus to different called numbers.

Don't Say I Didn't Warn You Department, Part Two: Your employees will also come up with dozens of additional uses for voice menus. Spare capacity and expandability are important issues.

6. *Zero Drop:* Does the system allow a caller to press 0 at any time and immediately switch to a live operator? (The answer is "yes" or this system shouldn't be on the market, but you're better off asking than having an unpleasant surprise.) Make sure the attendants attached to different numbers have different operators assigned.
7. *Rotary Phone Handling:* Can the system interpret rotary phone pulses as well as touchtones? A large number of private residences still use rotary phones. If calls from this demographic group are important to your business, an autoattendant that can't handle rotary phones may—no, will—result in loss of goodwill.

   If an autoattendant can't handle rotary phones, Rotary-to-DTMF convertors do exist. Explore this option in detail with your vendors. Make them responsible for this feature, however they choose to provide it to you. Otherwise, you could be in for a lot of mutual finger-pointing when something stops working.
8. *Voice Editing:* Do you have to record each announcement from beginning to end, or can you record "chunks" and assemble a complete announcement out of them? This feature can save a lot of re-recording for minor changes, especially when you're recording voice menus.

B. *Programming and Administration:* How you set up and keep track of everything. There isn't any point in asking for much detail in your RFP. Ask how much training you'll need.

C. *Backup and Recovery:* In case of different levels of system failure, you want to avoid re-recording announcements and recoding menus. Find out what provisions the system has for tape backup, and whether the backup system stores announcements, menus, or (preferably) both.

D. *Integration:* Your autoattendant will have to interface with your PBX, and possibly your ACD or VRU as well. Provide information about your PBX (and ACD or VRU as appropriate), and request a great deal of detail on this subject. Ask for a block diagram; an explanation of how the autoattendant signals other devices; and a description of any programming, parameter setting, or other techniques you will have to worry about to get the autoattendant to work in your current or projected environment.

II. Business Specifications
   A. *Warranty and Maintenance Terms:* Expect a one year warranty and maintenance after that for about 10 to 12 percent of the purchase price.
   B. *Payment Terms and Conditions:* You may negotiate a third up front, a third on installation, and a third on acceptance. Standalone autoattendants cost little enough that you simply pay for the thing.

Payoff Idea: Form a task force composed of an operator or receptionist from each area where you plan to install an autoattendant, and have them review your RFP and attend all sales presentations. Of everyone in your company, they will have the most insight into your telephone requirements. Take advantage of their expertise.

Even Bigger Payoff Idea: If you're planning on paying for your autoattendant with staff reductions, formalize a "no layoffs" policy, put it in writing, and publicize it heavily. Depending on your business you may achieve the same effect by:

- Reducing through attrition ("We won't lay you off, but we won't replace you when you leave, either")
- Redeployment ("We won't lay you off, but we may ask you to accept retraining for other assignments")
- Position enhancement ("Since your workload will drop, we will ask you to accept training for additional responsibilities")
- Growth avoidance ("Sales are increasing and so is call volume, but with the new technology we expect to handle the additional volume without adding more operators")

Evaluating Responses: See chapter 2, page 15.

INSTALLATION

Installation comes in two phases: *Physical installation* and *implementation of attendants.*

You have nothing to do with the physical installation except for scheduling a few hours when your phone system will be down. Most businesses install on Sundays to avoid interfering with normal operations, but for some (hospitals, for example) other times may be more suitable. Aside from this detail, simply stand back while your vendor bolts the box in place.

You have a lot to do with implementing the attendants. Here are your responsibilities:

1. **Identify all autoattendants.** Determine every area in your company that will want an automated attendant. If you have recently installed a new phone system, you probably have a users group already established (see chapter 3). As an alternative, use the task force you formed while developing requirements (see above) for this task.
2. **Plan each autoattendant menu.** Some areas may have no menu, just an announcement asking the caller to dial an extension, 0, or stay on the line for help. Others may want elaborate menu trees to direct callers to the right phone.

   Again, your users group or task force can do the brunt of the work. Your role is to help maintain some sort of stylistic consistency to avoid jarring callers as they go from one area to another.
3. **Script each menu announcement.** Produce a written script for each menu selection. Tips:
   A. Make all general announcements shorter than fifteen seconds.
   B. Include a *zero escape* in all announcements ("Press 0 and an operator will assist you"). You'll find some disagreement among experts as to when to present that option. Opinion: Say it last. Anyone who calls you more than once will know they can press 0, and it logically falls as a final option if you are presenting a menu.
   C. Make menu options shorter than five seconds.
   D. Make one person (you) responsible for writing all scripts. Have your users group, and perhaps your Marketing Department, review them.

E. Use sequential numbers for menu options rather than letter codes for mnemonics. In other words, make your choices 1 through 5. Don't try this: "For Sales, press S. For Accounting, press A. . ."
F. When presenting menu choices, mention the number *last:* "For Accounting, press 1. . ." rather than "Press 1 for Accounting. . ."
G. When timing announcements, speak slightly slower than normal. Don't keep announcements within their limits by speaking rapidly.
H. *Never* have more than three menu layers, and have three only under extraordinary circumstances. If all your announcements go to their maximum times, and if each menu has five choices, a caller will have spent a full minute dealing with menus by the time two menus have gone by.

4. **Code and Record Announcements and Menus.** Put everything in place. So far as coding everything, you may want to learn how to do this (it isn't especially difficult) or you may want to rely on your vendor. For actual recording, be objective about your voice. You may prefer to find someone else to do the announcing.

5. **Test Everything.** Your vendor will have tested the equipment and interface. Don't assume you've coded everything properly. Dial in and walk through every menu; press every button. Did you remember to test invalid entries?

Managing the System

You won't spend a lot of time managing your automated attendant, so we won't spend much time on the subject. You have two main concerns: Adding and changing announcements; and recovery from system failures.

Here's a surprise for you: From time to time you'll find yourself adding new announcements and menus for areas that discover the value of autoattendants after you've installed them elsewhere. (If your business is a retail operation, someone may discover that providing driving directions by phone is a good idea, for example.)

No problem, so long as your system isn't at capacity.

Here's another surprise: It is possible for your system to fail, either partially or completely. That's why you asked about tape backup in your RFP. **Urgent Recommendation:** As with all your other systems, store a current backup tape away from your primary place of business, and another one on the shelf in your office.

Another Recommendation: Keep a record of every script and menu in a notebook in your office, in case a backup tape turns out to be unreadable.

CHAPTER SEVEN

Voice Mail

VOICE MAIL, ALSO CALLED *VOICE MESSAGING* by those fond of syllable increases, is a nifty alternative to putting answering machines all over the place. And yes, it can make sense even if you already have a lot of answering machines, because voice mail can do a lot more, if your business learns to take advantage of it.

The *technology* of voice mail adds some slick features you'll never find in answering machines, such as message filing and a variety of ways to send messages. Don't be too optimistic—only a quarter of your employees will learn to use these at first.

More important than the technology of voice mail is the *culture* that comes with it. If establishing the culture doesn't interest you, the technology probably won't pay for itself—at best it will wash compared with putting answering machines where they're needed.

If your employees buy into the voice mail culture, they won't leave messages like, "Hi, this is John Smith. Please call me at extension 1234."

Instead, their messages will sound more like brief memos. "Hi, this is John Smith. I looked at the figures you sent, and except for an error in addition in the second column

Business Summary

Useful Life:	5 years
Cash Cost:	\$100–\$200 per phone system user
Amortized Cost:	\$25–\$50 per phone system user per year
Dedicated Staff:	One per 5,000 phone system users
Justification:	Reduces secretary/receptionist load, 'telephone tag,' and paper use. Replaces individual answering machines.

Cost Analysis (estimated)

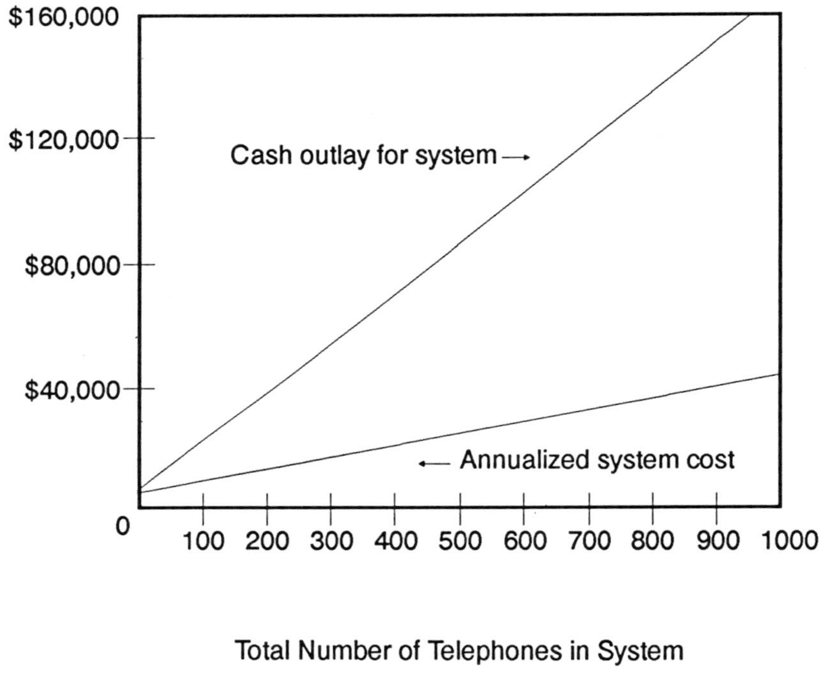

they look fine. If you need to talk, call me at 1234. Otherwise, print out a final copy and send it around."

We're all accustomed to griping about telephone tag, but few of us translate it into any action. If you compare those two scenarios, this should leap out at you: The second version accelerated a decision by at least one day.

As you go through the features described in the Requirements outline, think along the lines of improving communications, not simply taking messages.

If you have fewer than twenty employees, voice mail is probably overkill—use answering machines. With even twenty employees, you may find an economical voice mail system, or Centron voice mail, makes sense for your operation.

If you used an RFP for your PBX or key system, use one here. **Exception:** You may choose to simply buy from your phone system vendor, on the theory that your phone system and voice mail can't fail to work together, and if anything breaks you have only one number to call.

Few businesses have any requirements for voice mail in the sense that voice mail features are integral to their proper functioning. As mentioned earlier, voice mail is a technology that succeeds best if your business adapts to *it*.

As a result, a voice mail RFP is much closer to a one-size-fits-all affair than is true for most other technologies.

Requirements

I. Technological Specifications
  A. *Features:* What can the system do for your users? Think "mail" and you'll get most of what you should be looking for.
    1. *Message Recording:* Ask for a walk-through of the process by which callers leave messages.
    2. *Simultaneous Messages:* Can more than one caller leave a message for the same recipient simultaneously? This is one of the advantages voice mail has over an answering machine.
    3. *Zero Drop:* Make sure callers can get out of voice mail by pressing 0 to reach a live operator.
    4. *Transfer:* Can a caller press some key combination to dial another extension without having to hang up and call back?
    5. *Caller Retrieval:* If you hear a call go to voice mail, can you get it back? If any systems allow this it would be surprising (an advantage answering machines have over voice mail) but it's worth asking for.
    6. *Message Waiting Notification:* If you have new messages in your voice mail box, will the system (A) light your message waiting lamp; (B) produce a dis-

tinctive tone when you pick up the handset; or (C) none of the above (in order of preference).

7. *Playback Options:* The system lists all messages. It should let you:
   a. Speed up and slow down playback
   b. Replay the entire message
   c. Replay the last few seconds of a message
   d. Skip the message
   e. Delete the message
8. *Send Messages:* The process of recording a message and sending it to another individual through the system, without ringing the person's phone. Desirable options include sending messages to:
   a. Individual mailboxes
   b. List of numbers (in other words, more than one mailbox)
   c. Stored list (a previously defined distribution list, identified by some code. Stored lists come in two flavors, public and private. Ask whether only the system administrator can create public lists, or whether others can do so by definition of an appropriate privilege class.)
9. *Send by Name:* Will the system allow you to address messages by name? In other words, can you press telephone keys whose letter codes correspond to names, and have the system recognize them?
10. *Announcements:* What you hear before the beep that prompts you to leave a message. Ask for these useful features.
    a. *Personalized Greeting:* The system should allow users to record a personal greeting. Ask if the personal greeting comes before, after, or instead of the system announcement. Preferences differ, but personal greetings ought to replace the system announcement.
    b. *Default System Announcement:* The system should produce an announcement even if a user hasn't recorded a personal greeting.
    c. *Multiple Announcement Management:* The system should allow users to record different messages

for different situations, and switch among them easily. For example, you may store one message for days you are in your office and another for days when you're traveling.

Even Better: Some systems will automatically provide one greeting to inside callers and another to outside callers.

Another option: Providing one message when your phone is busy, another when you don't answer it.

d. *Autoattendant Features:* Many voice mail systems provide an automated attendant. See the previous chapter for details on what to look for.

11. *Message Forwarding:* The ability to forward a message you've received to another individual, with your spoken comments up front.
12. *Message Answering:* The ability to record and send a response to the sender of a message.
13. *Message Filing:* The ability to establish one or more permanent files for storing messages by category.
14. *Return Receipt:* The ability to flag a message so the system tells you when the recipient plays it back.
15. *Voice Forms:* The ability to prompt for specific information in a message, and record just the responses. For example, some banks allow customers to transfer funds among accounts using voice forms—the system prompts for the name, account number from which funds should be withdrawn, and the account number to which funds should be deposited. A spoken prompt precedes each piece of information, rather than using a single prompt to ask for everything.
16. *Pager Notification:* Having the system dial your pager when a message appears in your mailbox.
17. *Fax Store-and-Forward:* This is the "latest and greatest." Some voice mail systems now allow you to store incoming facsimile transmissions in a user's voice mailbox. Users can direct their faxes to the most convenient fax machine at their leisure.

B. *Security:* Few organizations need any level of sophisticated security, and nearly every system provides passwords. Ask about security so you understand what you get, but unless your organization is highly conscious of security, don't make too much of it.

Make sure security doesn't interfere with convenient use of the system.

Except for One Thing: If the system you select offers the ability to transfer out of voice mail, make sure you can prevent transfers to outside numbers. Otherwise, an outside caller could call a number and transfer from an employee's voice mailbox to a number in Mozambique.

C. *Programming and Administration:* Don't ask for a lot of detail in your RFP, unless you're prepared to read system manuals from every vendor you solicit. Ask how much training you'll need and how long it takes to set up a voice mailbox.

Very Important Question: Ask whether (and how) you can administer the system from a data terminal. Distribution lists are especially important, since administering the system from a spoken list quickly becomes impossible.

D. *Backup and Recovery:* Find out the system's provisions for tape backup, and whether the backup system stores announcements, messages, programming, or (preferably) all of them. Key questions: Are there any circumstances under which your users may have to re-record their announcements? Are there any circumstances under which mailboxes may lose messages?

E. *Integration:* Describe your phone system, and ask how each vendor's voice mail system will tie into it. Ask for block diagrams and a complete description of how calls switch into and back out of voice mail, and how the systems pass call information back and forth as well. In a fully-integrated system, your phone system will pass the name of the caller (for internal calls), type of transfer (busy, no answer, send all calls, etc.), and nature of call (internal or external), to allow your voice mail system to properly process the call.

Also, make sure each voice mail system will be able to light the message waiting light on user phones,

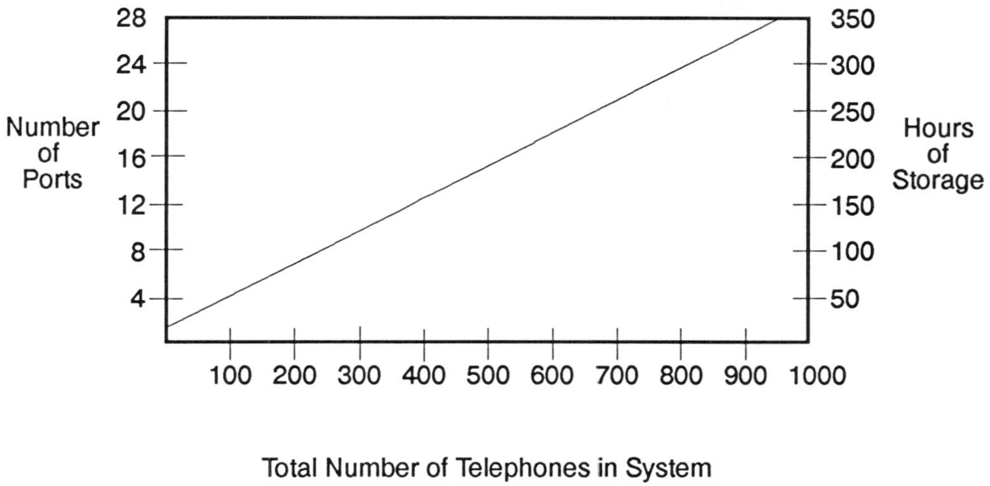

Figure 10.
Voice Mail Capacity Planning Chart

if your phones are of recent enough vintage to have them.

F. *Capacity and Expandability:* Describe your company, the uses to which it will put voice mail, and the number of employees who will use it (voice mail may not make sense for everyone). Ask your vendors for recommendations on the number of voice mail ports (simultaneous calls) and storage (number and length of messages).

Also, ask how expandable the systems are, and whether they can integrate the operation of more than one system. You need this information in case your initial size estimate is wrong or your organization grows.

Rules of Thumb Department: Plan on a port for every thirty to fifty employees, but never fewer than three ports (see Figure 10). If you have fewer than ten employees, voice mail rarely makes sense—use answering machines instead.

For storage, plan on about twenty minutes of storage per employee.

II. **Business Specifications**
   A. *Warranty and Maintenance Terms:* As with most technologies in telephony, a year is standard. Ask anyway. Ten to 12 percent of the purchase price is typical for maintenance contracts.
   B. *Payment Terms and Conditions:* Expect a third up front, a third on installation, and a third on acceptance.

Payoff Idea: You've heard this before by now. Form a group of key voice systems users if you haven't already, and involve them throughout the decision process. They'll be invaluable to you, both in helping select a system, and in making sure implementation goes smoothly. Especially with voice mail, where success depends on a change in how employees interact with the telephone, you need this sort of group.

Evaluating Responses: Review chapter 2 for suggestions on evaluating responses to RFPs.

Installation

As with most voice technologies, you won't have much to do in terms of physically installing a voice mail system. Your vendor will handle that. Unlike a PBX, where the changeover disrupts business, your voice mail system can go in without anyone noticing.

What to Expect from Your Vendor

Expect your vendor to deliver equipment, test it, and attach it to your existing phone system (if you use a different vendor to service your phone system, let the two vendors figure out who is responsible for what).

Don't expect a project plan for your voice mail system—the process is too simple to warrant one. Your vendor should provide two dates: When the equipment will arrive, and when installation will be complete.

Your vendor will also provide training, but you are responsible for scheduling it. You also have some responsibility for the content of that training (covered later in this section).

WHAT IS EXPECTED OF YOU

You are responsible for bringing voice mail into your organization. Nothing about this process is especially difficult—what follows may seem obvious. We'll cover both the obvious and more subtle points, to make sure you don't miss any.

You Have Four Key Tasks in Front of You: Deciding how long to retain messages; determining how each employee will use voice mail; ensuring every employee receives training in the mechanics of the system; and fostering the cultural change that will make voice mail succeed in your organization.

Retention: Don't make this decision on your own. Ask your user group to share in it. Your options are simple and obvious—one week or two. A week is enough for normal business, but two weeks lets someone take a longer vacation without losing messages. The tradeoff: The longer your retention, the more storage you'll need.

A Key Point: If you want one week of retention, set your system for ten days (to include a full week plus the weekends at both ends and the full first business day after). For two weeks, retention should be seventeen days (same logic).

If two-week vacations are common in your organization and you can afford the capacity, set retention at seventeen days.

Employee Requirements: You have some concrete detail work in front of you, because your voice mail users now have more complicated call coverage paths. Three strategies are common; you will have individuals in your organization who need each. First, you may simply assign a voice mailbox and have done with it—all calls roll over to the mailbox if nobody answers the phone.

Second, you may assign another individual—a secretary, receptionist, or even another individual with similar responsibilities—for primary coverage, and only roll over to voice mail if nobody answers either phone. (Don't go to three, because callers will hang up before their call finally stops ringing and reaches a mailbox.)

Finally, you may assign a voice mailbox for primary coverage but assign a zero drop to a secretary, receptionist, or what have you, if a caller needs to speak to a living human being.

Payoff Idea: Your phone system may let you route outside calls using one strategy, and inside calls using a different one. If so, some of your employees will probably want outside calls to roll to their secretaries while inside callers can go directly to their mailboxes. (You won't find this feature listed under Requirements, because it is a PBX feature, not part of your voice mail system.)

Figure 11 is a sample form you may use to keep track of each employee's requirements and training status. It's mostly self-explanatory, but let's make sure everything is entirely clear.

You may choose to organize your user list by department, or simply list all your employees alphabetically. Neither has any special benefits over the other—do whichever is most convenient.

Write in each user's name, last name first, in the first column. In the second column list each extension.

Inside Cover is the number to transfer internal calls to if the user's phone is busy or not answered within three rings. Write in "VM" if voice mail is the immediate coverage. *Inside 0-Drop* is the number to transfer internal calls to if the caller presses 0 during the user's greeting—usually the same as the inside cover, unless the coverage goes directly to voice mail.

Outside Cover and *0-Drop* describe how to route calls from outside callers. Enter these as you did for inside cover and 0-drop.

Use the column headed *Training Schedule* to keep track of when you have scheduled each user for training. Enter a check mark in the column headed *Done* when the user has completed training.

Training: Your vendor will provide training for all your employees. The training materials will cover the mechanics of using voice mail. Before training begins, meet with the trainers to make sure they are familiar with how your organization is set up.

You are responsible for establishing a training room with enough phones to allow one for every employee. (No, not all at the same time. Plan on about twenty employees per training session.)

The most difficult part of training is making sure everyone attends. As with your phone system, the higher up in management employees are, the less likely they are to go to

Voice Mail User Plan Department #______________ Page ____

User Name	Ext	Inside Cover	Inside 0-Drop	Outside Cover	Outside 0-Drop	Training Schedule	Done

Figure 11.
Voice Mail Planning

training. Nobody knows why this is. The most likely explanation: "I hire people to do demeaning things like this." Go figure.

Employees accustomed to answering machines may not understand why they have to log in and provide a password, and since they're used to having labeled buttons for every function, the process of listening to voice prompts and responding to them may be baffling. (Some employees may never progress to the point of entering responses before voice prompts finish, through an instinctive sense of politeness—even though it is a machine talking, they won't interrupt.)

Payoff Idea: If you have the authority, don't provide voice mailboxes to any employee—including the president—until that employee has completed training. If you don't, keep track and pester the living daylights out of everyone who hasn't gone through it.

Changing the Culture: Because making voice mail successful involves a cultural change, you are responsible for fostering that change. It can be done. It isn't even especially hard. It is time-consuming, trying, and requires a lot of patience.

You won't find any magic formula in this book, and you may have a better approach than these. A few suggestions you may find helpful:

1. **Don't call it a cultural change.** Call it a change of habit, or an adaptation to changing technology, or whatever. Make it seem obvious, minor, and non-threatening.
2. **Repeat the message often.** It's more than just an answering machine. Make presentations before, during, and after installation that emphasize how voice mail can improve communications. Make sure all employees remember they can use distribution lists and forward messages. Beat them over their heads and shoulders about leaving messages instead of "call me" notes.
3. **Include it in training.** Ask your vendor's trainers to help you. They can teach how to take advantage of voice mail, as well as the mechanics of its use.
4. **Secretaries are the key.** They'll train their bosses.
5. **Practice what you preach.** Well, this is obvious, isn't it? If you tell everyone how wonderful it all will be, and then call someone and say, "This is Ralph Cramden. Call me at 1234," you won't impress anyone. Leave the same kind of detailed

messages you want others to leave for you. Set a good example.

Remember to use the fancy features yourself. Your use of them will remind message recipients of the features.

6. **Use your company newsletter.** Put in an occasional tip on using voice mail.

Information-Only Mailboxes: Find out if you need any information-only mailboxes. Since most voice mail systems can double as automated attendants, some areas within your company may wish to use voice mail to provide information. Human Resources may wish to provide information about the company health plan; Promotion may wish to provide driving directions and company hours to customers. . . you have a lot of possibilities. Make sure your key users look for opportunities.

A Big Word of Warning: Here's a guarantee. . . you will receive a *lot* of complaints that "your damned voice mail system doesn't work right." The problem: Many employees won't think through how they want calls routed until after you've given them their voice mailboxes. Don't blame them for this. Few people can visualize (audiolize?) voice mail in operation until after they've used it for a month or so. Once they have it they can spot what they don't like.

Payoff Idea about the Big Word of Warning: Early and often, make sure your employees *expect* to want changes. Explain that all organizations experience a period of adjustment after installing voice mail, that it is normal, and that you want everyone to let you know how you can improve call routing.

Managing a Voice Mail System

Managing voice mail is neither difficult nor time consuming. As with your phone system (or anything else, for that matter), your key is to ask, "What do I need to know?"

The less pleasant version of the question is, "What terrifies me most that I could keep from going wrong?"

With voice mail, your tasks are pretty obvious. You need to:

1. Add new employees to the system

2. Delete departing employees from the system
3. Manage distribution lists
4. Keep information-only mailboxes up-to-date
5. Make sure you don't run out of space
6. Make sure you don't run out of ports
7. Protect yourself from system failure

New Employees: You can ask your Human Resources Department to notify you whenever a new employee starts work, or you can wait until a manager notifies you that new employees will be starting, and they'll need phones. **Recommendation:** Rely on the manager.

Payoff Idea: If your company has an orientation program for new employees, make sure it includes voice mail training. (You've already made sure it includes telephone training, haven't you? Add voice mail to your telephone course.)

As long as you're providing continuing training, make sure all employees are aware of it. Refresher training is a good idea, since an employee may not be prepared to learn all features of voice mail in the first session.

Departed Employees: You have two choices again—you can ask Human Resources to notify you whenever employees leave the company, or you can rely on their managers. **Recommendation:** Rely on Human Resources, the rumor mill, and notices for going-away parties.

Wait a day longer than your system retention period before eliminating the employee's mailbox. That way, nobody can complain that you have wiped out a vital message.

Managing Distribution Lists: You'll set up quite a few public distribution lists, such as "All Employees," "All Managers," and "Company Executives." Make sure you keep them up to date as you process adds, moves, and changes.

Information-Only Mailboxes: When you set up information-only mailboxes, ask whoever generates the information to give you an expiration date. Put that date on your appointment calendar, and when you reach it, contact the owner, either for updated information, for agreement to continue the same information (get a new expiration date in either case) or for agreement that it's time to delete the mailbox.

Managing Space: The philosopher Spinoza proposed that nature abhors a vacuum. He was wrong, of course—most of nature *is* a vacuum—but for human nature he was right on the money. No matter how much storage you provide, your employees will find a way to fill it up.

There are no magic formulas for managing space. Check it at least once a week. Making it part of your daily routine is even better. When you hit 80 percent in use, send a message around asking employees to clean up their old messages.

If you find your system is always full, you probably need to add storage. Before you do, ask your vendor to run diagnostics on the system to verify the space is in use. Defective hardware—especially a deteriorating disk—could make storage unavailable to your users.

Managing Ports: Your system should provide you with statistics on how many ports it uses during different times of day. Take advantage of these. If you have more ports than you need, take some out of service. You can either sell them on the used equipment market or keep them on the shelf for spares. Either way, you'll save ports on your phone system as well.

If you find times when all ports are in use, you'll need to dig a bit. Several factors could cause you to exhaust your port capacity:

1. **Too Small a System:** You could have simply sized your system too small from the beginning. If so, you'll need to add ports.

2. **Autoattendant Provided by Voice Mail:** Voice mail can provide autoattendant features to your phone system. If you're using these features, you may have used more of your voice mail capacity than you expected. If this is the case, consider buying a standalone automated attendant to free up your voice mail system.

3. **Phone System Problems:** Phone systems sometimes "busy-out" lines for no obvious reason. Yours may have done so, tying up voice mail ports indefinitely. If this happens often, the interface between your voice mail and phone systems may be subtly wrong. Have your vendor check it, and don't let him off the hook until you have a satisfactory answer. Don't allow this to become business as usual.

4. **Hardware Failure:** Ports can simply fail, and can do so in a seemingly infinite number of ways. Not all are obvious. If you

have ports busy at odd times of day, or some ports are always in use, have your vendor test them thoroughly.

Recovery from System Failure: As with every other system you have, back up everything to tape on a daily basis, keep a week's worth of tapes in rotation, and store at least one recent tape off site, just in case.

CHAPTER EIGHT

Universal Call Distributors and Automated Call Distributors

AUTOMATED CALL DISTRIBUTORS (ACDs), and their less expensive cousins, *Universal Call Distributors* (UCDs), are the heart of any call center. Both allow you to accept more calls than you have agents to handle, stacking them up until agents become available.

The difference between ACDs and UCDs is the latter deals out calls in a fixed order, so you have a "hot seat," while the former allocates calls more evenly. Also, ACDs provide a wealth of real-time management information absent from UCDs.

Decide which one you want based on your requirements. If your phone room has even five agents, you probably need an ACD.

Every reader of this book has dealt with an ACD at one time or another. Probably, you have dealt with enough of them—buying tickets, asking for customer service, or stopping your newspaper for the weekend—to recognize when companies use ACDs to improve service, and when they use them to cut payroll costs.

Few readers will have the familiarity with this technology that they do with basic telephony or voice mail systems,

Business Summary

Useful Life:	3 years
Cash Cost:	$200–$300 per seat
Amortized Cost:	$150–$250 per seat per year
Dedicated Staff:	One quarter-time (or less) administrator
Justification:	Reduces call center staffing by load-balancing during peak periods. Increases customer satisfaction.

Cost Analysis (estimated)

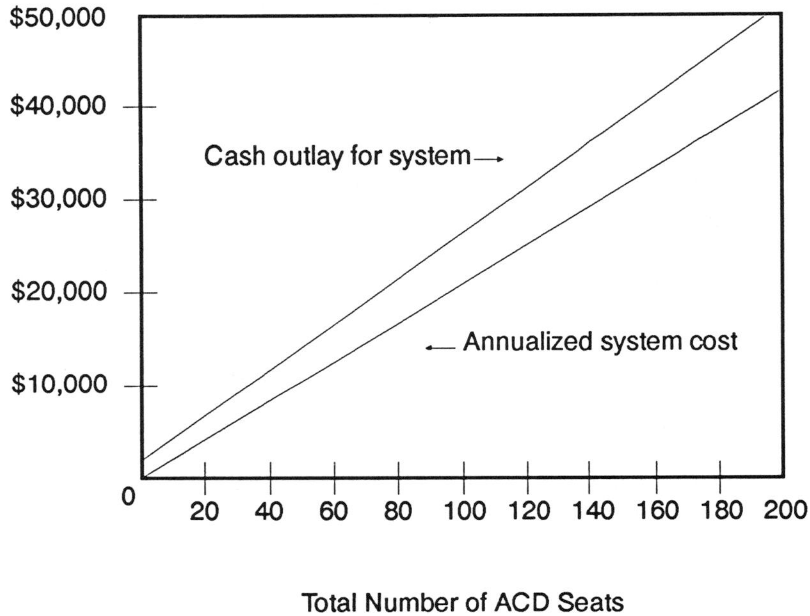

unless they already have an ACD in use. Regardless, this section will take you from start to finish.

WHY A THREE-YEAR LIFE?

At the beginning of this chapter, you saw an estimate of three-years as the useful life of an ACD. Since ACDs use the same basic technology as PBXs, you may find this estimate questionable.

An ACD you install today will run fine for at least seven years, just like a PBX. That's not the issue.

You buy an ACD for a competitive edge (or more likely, to blunt your competitor's edge, since your ACD just helps you stay even). In three years, new call processing technology will let you and your competitor process calls much more efficiently than you can with a simple ACD. If you don't want to lose your edge (or let your competitor gain one) you'll have to upgrade or replace your ACD to take advantage of the features that will appear in the next few years.

Chapter 10, Interactive Voice Response Systems, deals with these new technologies in more detail, as does the final chapter.

Your Recommended Goal: Use ACDs for short-term load-balancing, not as expensive hold buttons. Staff for the hottest five-minute period in each hour and you'll rarely ask a customer to hold for more than two and a half minutes, never more than five minutes. (Don't take these numbers too seriously. Everything depends on the nature of your specific business.)

By the way. . . ACDs are just as useful for slow periods. You may provide twenty-four-hour service even though you only get a handful of calls during the midnight-to-six shift. With an ACD you can hire a single night agent and still cover the occasional time when you have more than one caller. You even cover "biology breaks" without exposing callers to a busy signal.

ACDs aren't a new idea, and they're designed to solve a particular set of problems. One approach is to forgo any requirements analysis. In this variation, you simply decide you have that particular set of problems. Call several companies out of the Yellow Pages and talk to their salesmen. Do business with the one who makes the most sense.

Or, you may simply buy from the supplier of your phone system, because then your phone system and ACD can't fail to work together (oh yeah?), and if anything breaks you have only one number to call.

Or, go ahead and write a formal RFP. No matter which approach you choose, you have some important questions to ask. ACDs may be commonplace, but they aren't commodity items.

ACDs can be specially-programmed PBXs, or adjuncts to an existing PBX that add special switching logic. It shouldn't be too surprising that ACD requirements match PBX requirements quite closely.

As usual, use the requirements listed below as you need them. Call centers tend to resemble each other, but no two businesses are exactly alike.

Payoff Idea: Be open to modifying your call center operations, or restructuring your call centers entirely. You may find that a new ACD allows you to create a more efficient organization by consolidating some functions you now keep separate. Keep an eye out for these opportunities as you evaluate RFP responses.

Another Payoff Idea: Use your ACD instead of an autoattendant for your main switchboard. When call volume permits, your operators will handle calls directly. Program your ACD to pick up after three rings, so calls don't go unanswered.

Another Notion: Read chapter 10, Interactive Voice Response Systems, before you write your RFP. It may make sense to buy an ACD with a built-in IVR system.

A Key Decision: You may have the option of augmenting your existing PBX with ACD software or an adjunct processor (a box on the side of the box that takes care of some of the work). If this option seems attractive (you should save a bunch by going this route) make sure your PBX has enough spare capacity to absorb the load. How to make sure? You *must* use the same vendor you used for your initial PBX. Have the vendor certify your capacity and accept contractual responsibility for the estimate. It may be wrong, but at least you won't be out of pocket for the result.

Requirements

I. Technical Specifications
   A. *Features:* What can the system do?
      1. *Group Definition:* Small ACDs distribute calls within a single group of call handlers. Larger ones let you define more than one group. Decide which one you need.

2. *Call Distribution:* What you're buying the system for. Ask for details on how it allocates calls to agents. Especially, ask about:
   a. *Overflow groups:* You should be able to configure the system so excess calls from one group overflow to another. You may want this to be immediate. More likely, you will want this to happen when service degrades to unacceptable levels (when callers have been waiting for a few minutes).
   b. *Time-of-day call direction:* Unless you have a twenty-four-hour-per-day call center, you will want to route night calls differently than calls during business hours—possibly to a voice mailbox.
3. *Autoattendant Features:* See chapter 6, Automated Attendants, for details. Important if you want to use your ACD in place of an autoattendant, or if you want callers who are waiting on an ACD to have the option of branching to other areas of your company.
4. *Music On Hold:* Instead of dead air, provide callers on hold with something to listen to. Look for the ability to provide different recordings to each group. Ask whether the system treats recordings as announcers within the ACD, or whether you would use an external tape loop to provide these, and why the manufacturer chose the strategy it did. (Neither has a clear advantage.)

   Payoff Idea: Instead of music, use hold time as an opportunity to present special promotions, information on new or unique products, or general themes about how wonderful your company is.

   Music is dangerous. Classical music will bore rock-and-rollers, rock-and-roll will offend those attuned to elevator music, and elevator music will send most callers on an astral journey to Xanadu.
5. *Group Administration:* How you assign agents (that is, your call center personnel) to groups. Ask if the system lets you assign a station to more than one group (you may have some agents trained in more than one operation). Also ask who can move sta-

tions from one group to another (supervisors or only administrators?) and whether this is a manual operation, a timed, automated operation, or whether it can be done either way.

Make sure you can administer the system from a computer terminal or (better) a personal computer.

6. *Messaging:* Lets supervisors display a message on an agent's telephone display. Example: "Finish already!"
7. *Broadcast:* Lets supervisors display a message on all agents' displays. Example: "Lots of calls waiting."
8. *Internal Transfer:* Allows a supervisor to transfer calls to specific agents. Useful for handling personal calls, or follow-up calls from customers who want to speak to a particular agent.
9. *Call Monitoring:* Lets supervisors listen in on calls without the caller or agent knowing it (the term is "Silent Monitor"). This process can cause resentment, but you have little choice. Even though it smacks of Big Brother, your customers come first, and you have no other way of ensuring they receive quality service.

   Payoff Idea: Make sure employees know you'll be monitoring calls, and are comfortable with the reason. If they aren't, ask them to provide alternatives that ensure the same goal. Make them your allies in the process, not adversaries.
10. *Scheduling:* Different systems provide differing levels of support for maintaining schedules and tracking actual attendance against the schedule. The fanciest systems predict call volume on an hourly basis. Using predicted volumes and historical data regarding individual productivity statistics, they help you develop optimum work schedules.
11. *Real-Time Reporting:* Display of current call statistics for call center supervisors. Key performance measures:
    a. Number of calls waiting
    b. Average wait time
    c. Wait time for oldest call

d. Average call time (summary and by call center agent)
e. Number and percent of calls lost.
f. Agent status (number of agents logged in, number available, number on break, number involved in after-call work)

The first three measures cover calls currently in the queue. The last two are taken over an interval somewhere between five and fifteen minutes.

A Heck of a Lot More Useful than You'd Expect: Color graphics displays. They let you display the above measures graphically, and compare them to predefined levels of service.

Example: You may set a service level of twenty seconds for average wait time. A graphic display shows average wait time as a bar graph. A color graphic display shows the bar in green when the measure is below twenty seconds and in red when it exceeds twenty seconds. A supervisor can monitor statistics from across the room with one of these puppies.

Pretty Useful, Probably: A status signboard, to let all agents see key statistics. If your agents don't know they have a lot of waiting (and increasingly impatient) callers, they can't respond appropriately.

12. *Management Reporting:* The ability to print out various call center statistics for different time intervals. Look for a good set of standard reports providing call center and individual agent performance (the same measures for which you need real-time data—see above).

Also look for the ability to print out group configuration information, including intergroup call routing information (overflow groups and so forth).

Very Important: Make sure the system provides reports through database techniques, and ask about:

a. *Database retention:* How long records are kept in the system. Two-year retention lets you make year-to-year comparisons, and improves the system's ability to forecast call volume.

b. *On-demand reports:* Your ability to print whatever report you want, when you want it, for any interval you choose. The alternative is to print month-end reports and store the printouts forever, because you can't re-create them.

c. *Custom reporting:* Whether you can create your own custom reports and customize those provided with the system (you don't want to pay for an outside programmer to do this).

d. *Data export:* Whether the system allows you to export data into a database or electronic spreadsheet program on a PC.

B. *Station Features:* Some features are dependent on the telephones (stations) placed on agents' desks. Good ones to ask about:

1. *Ergonomic Headsets:* You *must* provide headsets to your agents for handsfree operation. Don't even think about shoulder cradles—you'll lose so much money to workers compensation claims from repetitive stress injuries, you won't be able to stay in business.

2. *LCD Display:* Not a negotiable item. An ACD that doesn't provide at least a one line LCD display limits your options tremendously.

   Excellent New Trend: Providing station features from within a personal computer. A card inside the PC accepts a headset jack.

3. *Manual or Automated Pickup:* Lets agents choose whether to press a button to accept calls, or whether to provide a "ziptone" (a beep) through the headset as a warning, and then automatically connect the call.

4. *Multiple Groups:* Assuming your ACD can assign a station to multiple groups (see above), the agent's station must indicate the group for which an incoming call is intended.

5. *Standard Call Functions:* Provide hold, transfer, conference, and speed dialing buttons.

6. *Station Login:* Agents must be able to log in and out so the system knows to whom it may route calls.
7. *Agent Available/Unavailable:* A pair of buttons that lets an agent take a break. You want an agent to take a washroom break without having to logout.
8. *After-Call Work:* Another button that lets agents indicate that they are unavailable because they're completing a transaction (some calls may require paperwork after the caller has hung up).
9. *Multiple Headset Jacks:* A jack on both sides of the phone. This accommodates right- and left-handers, and lets a supervisor plug in to listen to a call from the agent's desk.
10. *Call Waiting Light:* A light that tells agents that calls are waiting in the queue. You may want agents to provide maximum service during slow times, but to hurry calls along when things get busy. A call waiting light lets them know.

C. *Security:* Security isn't a big deal for an ACD, but if you have a call center of any size you have some requirements. You'll want supervisor logins hidden behind passwords, because ACDs store employee performance information. Obviously, your administrator login must be secure, since it allows you to change the system configuration.

D. *Programming and Administration:* Don't ask for a lot of detail in your RFP, unless you're prepared to read system manuals from every vendor you solicit. Ask how much training you'll need, and how long it takes to set up a typical ACD group.

E. *Backup and Recovery:* Find out the system's provisions for tape backup, and whether the backup system stores announcements, programming, or (preferably) both. Your key questions: Are there any circumstances under which you may have to re-record your announcements; and are there any circumstances under which you'll have to manually reprogram the system.

   Few systems now provide for tape backup of announcements, although this is changing. As an alternative, ask how to dub announcements from tape. If this process is simple enough, record to tape first and store your original tapes for announcement backup.

F. *System Architecture and Integration:* Describe your phone system, and ask how each vendor's ACD will tie into it. Ask for block diagrams and a complete description of how calls switch into and back out of the ACD, and how the systems pass call information back and forth as well. In a fully integrated system, your call center agents will be able to transfer and conference calls into and out of the ACD with no restrictions beyond those experienced by any other telephone user.

Also ask how each vendor's ACD would tie into an interactive voice response unit (covered in chapter 10).

Very Important: Depending on your vendor and existing phone system, your ACD may be a standalone system (a special purpose PBX), an adjunct processor (a box that bolts to the side of your existing box), or software that runs in your existing PBX. Each has different ramifications for tying into your existing telephone environment, and for the flexibility you'll have in dealing with special situations.

Ask which aspects of call routing the ACD handles by itself, and which ones it hands off to your PBX. You need to make sure you'll get features such as time-of-day call direction with your specific configuration.

Similarly, the ACD may provide announcements by itself, or by a separate piece of equipment that attaches to it. Advantage to the former: One less interface to break. Advantage to the latter: You can select (and upgrade) your announcers without affecting your ACD.

Payoff Idea: Describe your data processing environment, and ask what provisions the manufacturers have made for integrating their ACDs with the data processing side of your call center operation, and whether they provide an *open architecture.* (Open architecture means the manufacturers have provided detailed instructions on how an outside computer can obtain call information from their system, and control call switching. Chapter 10 also discusses this topic.)

As Usual: You've read this before, but it's worth restating. Don't let your vendors snow you with gobbledygook. If they use incomprehensible jargon, persist in asking for clarifications, just as you would with a waiter in a French restaurant. In a French restaurant, you want to know you'll be eating snails *before* they're delivered from the kitchen. With your ACD you want to make sure it ties into your existing telephone system seamlessly.

G. *Capacity and Expandability:* Describe your company, your call center(s), and the number of employees involved. Ask your vendor for recommendations on how much trunking and announcer capacity (number of announcements and total message storage) you'll need.

Also, ask how expandable their systems are in terms of number of groups, stations, trunks, and announcements, and whether they can integrate the operation of more than one system. You need this information in case your initial size estimate is wrong or your organization grows.

Very Important Questions:

1. Can you flexibly allocate announcement time among all announcements, or does each announcement receive a fixed amount of time?
2. Does the system provide a fixed *number* of announcements, or simply a total amount of recording time you can allocate as you please?
3. Can you easily reconfigure the system yourself, without major expense? If you want to reorganize your call centers, add or remove stations, or change the way you handle calls, you don't want your ACD to inhibit your decision.

Rules of Thumb Department: Typical call centers operate well with a 2:1 trunking ratio (two outside trunks for each call center agent). For each group, plan on at least four announcements (first announcement, repeat announcement, an after-hours announcement, and one more for good luck).

Payoff Idea: Buy enough capacity for several internal groups. You may have an internal Help Desk, for example, or an Employee Benefits Information Line. The incremental cost of additional capacity is small compared with the benefits you'll experience.

II. Business Specifications

A. *Warranty and Maintenance Terms:* Expect the standard one year warranty. Ten to 12 percent of the purchase price is typical for maintenance contracts, but ask.

For your telephone system vendor only: Ask if one customer support department handles all products. If separate help lines cover your PBX and ACD, when you have a problem you could...no, you *will* have each group blaming the other's product.

B. *Payment Terms and Conditions:* Typical terms are a third up front, a third on installation, and a third on acceptance. Ask.

C. *Training:* Ask how many training sessions each vendor will provide, the maximum class size, and the nature of any training materials.

Expect your vendor to provide training for every call center agent you have, and separate training for supervisors. Anything less is not acceptable, unless you get a tremendous discount. If you do, use it to hire a trainer.

Also ask about administrator training—how many administrators you can train, how much training they'll need, and whether training is local or out of town.

Payoff Idea: You've heard this before. Form a group of key users (call center supervisors and agents) and involve them throughout the decision process. They'll be invaluable to you, both in helping select a system, and in making sure implementation goes smoothly.

Evaluating Responses: Review chapter 2 for tips on evaluating responses to RFPs.

INSTALLATION

Installing an ACD is a lot like installing a phone system (depending on what you're buying, it may *be* installing a phone system). Read or reread the section on phone system installation, page 30, before going on here—the same basic rules apply.

Meet regularly with your key-user group. You can't fulfill any of your responsibilities for an ACD installation without their full cooperation. Make sure you keep them up to date on how the installation is progressing.

During the installation you need to:

1. Determine what groups you need to set up.
2. Find out what extensions go into each group.
3. Decide how to process calls (how long should they queue before rolling to an overflow group or supervisor, time-of-day changes, etc.).
4. Script and record all announcements.
5. Handle details such as what music or message you'll play to people waiting for agents.
6. Make sure all supervisors and agents (and you) receive system training the week before you cut over to your new system.

Payoff Idea: Whether you're planning to manage your ACD, or whether you're planning to contract out all management work, make sure you understand *in detail* how your vendor has engineered solutions to your requirements into the installation.

Make sure you know what processes your ACD handles and which ones it hands off to your phone system. Make sure you know how your phone system and ACD are physically interconnected, and the programming needed on each to configure the link. If your ACD uses an external box for announcements, make sure you know how it attaches to the ACD itself.

You need to know all these things first hand because your relationship with your vendor may deteriorate over time. The last thing you need is a vendor holding a gun to your head, which is what you'll have if only the vendor knows how your system is hooked up.

So take careful notes, make good diagrams, and put the result in your ACD project notebook.

MANAGING AN AUTOMATED CALL DISTRIBUTOR

There's almost nothing to it. The system handles all your recordkeeping, so you have little to keep track of on paper. New employees need training, but their supervisors will be responsible for that.

Payoff Idea: As with your other voice systems, you may contract with your vendor to manage the system. Key advantages: You only pay when you need service, and training is not your expense. Key disadvantage: You've given up some control. If you do outsource, schedule periodic records reviews to make sure your vendor is keeping proper track of things. Have the vendor explain how your system parameters, such as your group definitions, announcements, and special programming, are kept track of.

You will need to add stations to existing groups, reconfigure groups, re-record scripts, change music or announcements, and so forth, but all these activities are simply installations in miniature. You won't need to develop any special skills to handle these situations.

Your only significant responsibilities will be running nightly tape backups **(as usual, store at least one current tape off-site),** running and distributing reports, and managing your relationship with the vendor.

Speaking of off-site storage, maintain copies of all announcement scripts in a notebook, and keep that notebook in your off-site storage facility (which may be a home bookshelf).

If your system can't provide management reports on system configuration, you will need to maintain paper records showing all groups and their call handling parameters, including the handling of overflow groups. The forms shown in Figures 12 and 12a are all you need. Keep them in the same notebook as your scripts.

The ACD Group Definition Form will help you keep track of how you've defined each of your groups. Each line de-

ACD Group Definition Form					
Group	From	To	Message	Overflow Group	Overflow Condition

Figure 12.
ACD Group Definitions

ACD Station Assignment Form Group: _______

Station	From	To	Station	From	To	Station	From	To

Figure 12a.
ACD Station Assignment Form

scribes one group (first column), for a particular time of day *(From/To)*. In the column titled *Message,* refer to the script used for that entry.

In the next column, enter the name of the overflow group, if any. Under *Overflow Condition,* describe the conditions under which the system passes calls to an overflow group. Examples: "Wait > 20 sec" (oldest call has been waiting more than twenty seconds); or "> 30 in queue" (more than thirty calls are waiting in the queue). Enter whatever criteria you use for that group.

Use the ACD Station Assignment Form to keep track of which stations you've assigned to each group. Enter a Group ID at the top of each page. The form provides three columns. In the first, enter the station ID (extension). In the *From/To* columns, enter the time of day for which you have assigned the station to that group.

A Final Point: While your phone room supervisors and agents know the most about how they use the ACD, you know the technology. Don't assume that once you've installed the system it won't need changing. Keep an eye out for ideas—for new ways of routing calls, better ways of presenting messages, variable messages, and so on. Also, keep an eye on the monthly statistics, to see if anything unusual pops out. Meet with your customers (the people who use the ACD) on a regular basis to make sure you're making the best use of the system, and the information it provides.

After all, here's where your technology meets your company's customers.

CHAPTER NINE

Power Dialers and Outbound Automated Call Distributors

If you have a group that calls names from a large list, you probably need at least a power dialer, and probably an outbound automated call distributor. ACDs automate call center operations, and make your staff more effective by automatically routing incoming telephone calls. For groups that place a lot of calls, such as telemarketing groups, power dialers and outbound ACDs serve an equivalent function. Odds are, one will pay for itself in less than a year by increasing the effectiveness of every sales agent.

Even if you haven't heard the term *Power Dialer* before, you probably understand the idea. As the name implies, it dials the phone for you, much faster than you could yourself. You save wear and tear on your knuckles (not so trivial an issue if you have any concerns regarding repetitive stress injuries), shave five seconds or so from each call, and eliminate wrong numbers. If you run a small call center, a power dialer may be all you need. You can "homebrew" a power dialer using simple personal computers, modems, and any of a hundred off-the-shelf software packages. (You'll find explanations of the technology you need in the companion volume to this book. Refer to the sections on PCs, Local Area Networks, and modems in *Computer Systems for Every Business.)*

Business Summary

Useful Life:	3 years
Cash Cost:	$200–$300 per seat
Amortized Cost:	$150–$250 per seat per year
Dedicated Staff:	One quarter-time (or less) administrator
Justification:	Increases effectiveness of outbound call centers by eliminating dialing and ringing time (about 30 seconds/call)

Cost Analysis (estimated)

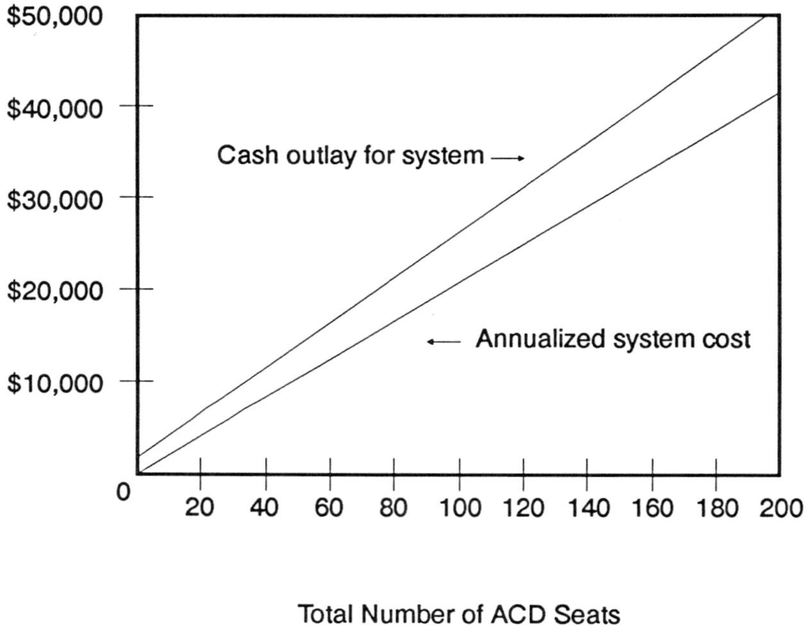

In this section we'll concentrate on full-blown outbound automated call distributors. The key technology for an outbound ACD is *predictive dialing*. The system monitors all agents and keeps statistics on average call duration and average connect time for newly-placed calls. When it predicts an agent will become available, it places a call to the next number on its dialing list.

If the wind is just right and Capricorn is rising, some hapless prospect will answer the phone just as an agent finishes talking to a previous lead. BANG! The system switches the new call to that agent, with no time wasted looking up the next number, dialing it, and waiting for someone to answer. Time saved depends on many factors, but thirty seconds per call is not uncommon. If your average call is five minutes long that represents a 10 percent gain in productivity.

Here's an obvious point: If you pay on straight commission, the productivity gain buys you very little—you're paying per sale. You'll save on employee benefits and on the number of call stations you have to equip (10 percent fewer sales agents), but we're not talking about a lot of money.

Payoff Idea: Restructure your commission package based on the new technology. You can give each sales agent a smaller commission for each sale and pocket the difference, while still giving them more income. In other words, split the productivity increase with your sales force.

Negative Payoff Idea: Some systems will play prerecorded messages. Don't do it. If you're tempted, ask yourself this: Would you like to receive calls from a machine?

Not Just for Telemarketing: You may find other uses for an outbound ACD. For example, if your organization surveys its potential customers, an outbound ACD may be just the thing you need (if surveys tend to be long, a simple power dialer may be sufficient). Or, you may find it helpful for collections.

Why a Three-Year Life?

Outbound ACDs have a three-year life for the same reason inbound ACDs have a three-year life. They'll keep on working long past that point, but you may find your shiny new system has become the telecommunications equivalent of a buggy whip—serviceable, but useless.

You're buying this to give you an advantage over your competitors. In three years, if you don't keep up, new call processing technology may let your competitors leapfrog you. If you don't want to lose your edge (or let your competitor gain one) you may have to upgrade or replace your system to take advantage of the features that will appear in the next few years.

You'll find two approaches to outbound ACD design—turnkey systems and open architectures. Turnkey systems provide "full call center solutions." In other words, they provide call handling, but also data-entry screens, commission reports, and everything else you need to run your call center.

Systems with open architectures may provide full-featured solutions as well, but they're more likely to provide only the call handling component of your operation. Instead of bundling in the rest, their designers make them easy to hook into your existing data processing setup, which you'll use for data-entry, database management, and reporting.

Which is better? Both. It all depends on your requirements.

Here's something new: A technology that affects your data-processing operation. As you develop your requirements (remember, the ones in this book are just your starting point) you'll have to make design decisions that depend on your existing data-processing systems.

If your sales agents already have a complete data-entry system they are familiar with, and that works well for your company, you may decide to go with an open architecture to preserve your existing systems.

If you don't have a system already, or if you have one but everyone agrees it just isn't adequate, you may prefer to buy a turnkey system instead.

A Key Decision: As with inbound ACDs, you may be able to enhance your existing PBX with outbound ACD software, or an adjunct processor that handles the computational load, but uses the PBX to take care of call switching. This option will probably seem attractive (it should cost a lot less) but make sure your PBX can deal with the additional call volume and processing load. Since you have no way of knowing for sure, you *must* use the same vendor you used for your initial PBX to certify your capacity, and accept contractual responsibility for the estimate. It may be wrong, but at least you won't be out of pocket for the result.

The Usual Payoff Idea: Put together a group of key users to help figure out your requirements. Then, keep them involved through the installation. Most companies buy outbound ACDs for a specific purpose and group—usually telemarketing, sometimes collections, and occasionally some other group. Your group of key users becomes the telemarketing manager, any supervisors in the area, and a few representative call agents (they're the only ones who *really* know what you need). Then add one or more people from your data-processing organization, and if you want to be radical, someone from your main list house.

Requirements

I. Technical Specifications
   A. *Features:* What can the system do?
      1. *Predictive Dialing:* The key component. Make sure the system does it (of course). Ask for performance statistics. Two key measures:
         a. *Percent connections put on hold:* With predictive dialing, the system will complete some calls before an agent becomes available. The lower the score, the better.
         b. *Average agent wait time:* You're buying predictive dialing to minimize this statistic, which tells you how long an agent waits between finishing one call and starting the next.
      2. *Integral Data-Entry:* Turnkey systems provide data-entry terminals, and some open systems may have them as an option. If you want a turnkey system, ask how easy it is to customize the data-entry environment.

         A critical question for turnkey systems: Ask what database management system (DBMS) they use. The only unacceptable answer is a custom file format. So long as the manufacturer programmed the system using a commercial DBMS, you should be safe.

         (Why? It's easy to get at the information in a commercial DBMS, so if you want to develop your own custom reports, or bring data into a PC so you can fiddle with it, you can. Also, commercial DBMSs are far more likely to have robust measures for preserving your data when your system fails, and to have utility programs for rebuilding files in case of disaster.)

         Absolutely Critical: You may not *want* integral data entry. If you have a satisfactory order-entry system (or whatever) you have no reason to change. Take that as an indication you should consider open architecture instead.

Whether you buy a turnkey system or one based on open architecture, you have the same goal: Your call agents should see a processing screen with all information already in place by the time they start talking.

3. *List Conversion:* Whether you're calling from a house list or one you're renting, you need some way of bringing list information into your outbound ACD. You have three separate issues to deal with—media, character set, and format. Research all three with your list houses and internal data processing group (for your house list) and present that information in your RFP.

   Media refers to the physical method of transporting data. The most common methods of distributing lists will be magnetic tape, 5¼″ floppy diskette, or 3½″ floppy diskette.

   If your list houses provide magnetic tape, ask your data-processing group to evaluate answers because of the wide variety of ways information can be put on tape. If you get floppy diskettes, don't worry. You can always stick a PC in the loop, and PCs can interface with *anything.* (Still, ask. It's better if you don't have to add the additional step.)

   You may have a list house that only provides information on printed labels. If so, you may want to look into *Optical Character Recognition* (OCR), or you may want to change list houses. You should probably change list houses.

   Character Set is an issue for your list houses, and your internal data-processing group. Any outbound ACD you're likely to buy will use something called ASCII (American Standard Code for Information Interchange). Your list house may be accustomed to providing something else called EBCDIC (Extended Binary Coded Decimal Interchange Code), especially if you use IBM mainframes. Make sure your list house can provide lists in ASCII.

Format refers to how you'll need to rearrange information from the way your list provider stores it to the way your outbound ACD needs to use it. Don't spend a lot of time asking your list houses for specifics on this one. Instead, ask for a detailed description of how the inbound ACD imports list data, and whether it can convert data from multiple formats or whether you'll have to do the reformatting yourself. (Having it as part of the ACD is better.)

4. *Messaging:* Lets supervisors display a message on an agent's telephone display. Example: "Finish already!"
5. *Broadcast:* Lets supervisors display a message on all agents' displays. Example: "You're above quota. Relax."
6. *Internal Transfer:* Allows a supervisor to transfer calls to specific agents. Useful for handling personal calls, or follow-up calls from customers who want to speak to a particular agent.
7. *Call Monitoring:* As with inbound ACDs, lets supervisors listen in on calls without the caller or agent knowing it (the term is "Silent Monitor"). Your customers come first, and you have no other way of ensuring your agents handle calls properly.

    Payoff Idea: Cast this positively, not punitively. Make sure employees know you'll be monitoring calls, and are comfortable with the reason. If they aren't, ask them to provide alternatives that ensure the same goal. Make them part of the decision, not opponents to it.
8. *Management Reporting:* The ability to print out various call center statistics for different time intervals. For turnkey systems only, look for a good set of standard reports providing call center and individual agent performance. (For open architecture systems, you'll continue to use the management reports you're accustomed to.)

    Very Important: Make sure the system provides reports through database techniques, and ask about:

    a. *Database retention:* How long records are kept in

the system. Two-year retention lets you make year-to-year comparisons.

b. *On-demand reports:* Your ability to print whatever report you want, when you want it, for any interval you choose. The alternative is to print month-end reports and store the printouts forever, because you can't re-create them.

c. *Custom reporting:* Lets you create your own custom reports and customize those provided with the system (you don't want to pay for an outside programmer to do this).

d. *Data export:* Lets you export data into a database or electronic spreadsheet program on a PC (or, for that matter, to your mainframe—one way of providing batch interfaces to mainframe systems).

B. *Station Features:* Some features are dependent on the telephones (stations) placed on agents' desks. Good ones to ask about:

1. *Ergonomic Headsets:* You *must* provide headsets to your agents for handsfree operation. Don't even think about shoulder cradles—you'll lose so much money to workers compensation claims from repetitive stress injuries, you won't be able to stay in business.
2. *LCD Display:* Not a negotiable item. Having this feature allows supervisors to discreetly signal your call agents without interfering with conversations.
3. *Manual or Automated Pickup:* Lets agents choose whether to press a button to accept calls, or whether to provide a "ziptone" (a beep) through the headset as a warning, and then automatically connect the call.

   You may not need this feature, since in an outbound environment you don't want the people you're calling to know a machine did the dialing.
4. *Standard Call Functions:* Provide hold, transfer, conference, and speed dialing buttons.
5. *Station Login:* Agents must be able to log in and out so the system knows to whom it may route calls, and who earns what commission (for turnkey systems).

6. *Agent Available/Unavailable:* A pair of buttons that lets an agent take a break. You want an agent to take a washroom break without having to logout altogether.
7. *After-Call Work:* Another button that lets agents indicate they can't take a call because they're completing a transaction (some calls may require paperwork after the caller has hung up).
8. *Multiple Headset Jacks:* A jack on both sides of the phone. This accommodates right- and left-handers, and lets a supervisor plug in to listen to a call from the agent's desk.

C. *Voice System Interfaces:* How your new system will hook up to your existing phone system. Describe your existing telephone system in some detail, so your vendors will know what they're up against. In your description, include the make, model, and age of your telephone system and inbound ACD (if you have one), number and type of central office lines or trunks, and number of telephones, modems, fax machines, and whatever else your system supports.

Even with turnkey systems you'll sometimes want to transfer calls, both in (even call center employees have families) and out (calls sometimes need referral to others in the company). Ask for *specifics* in terms of what a transfer (or conference) will look like to the agent. You want it to look the same for any employee with a telephone.

For open systems you have another question to ask, namely, can the vendor's system work with your existing PBX or key system, or do you need another one specifically for this purpose? As mentioned previously, you can achieve substantial economies by buying an open system and using it as an adjunct processor (a box on the side of the box, remember?), but only if the adjunct processor can work with your phone system.

Ask for a block diagram and a detailed description of how the outbound ACD will hook up to your existing phone system. Beware of generalities. Statements like, "...all we need is an RS-232 port to hook up to . . ." betray a dangerous lack of insight. Similarly, the phrase "We've never worked with your exact make and model, but it shouldn't present any special difficulties," should cause the hairs on the back of your neck to stand up, unless the word "because" follows it (". . . because as far as your system is concerned, our system looks just like a rotary hunt group of 2500 sets that use standard touchtone signals to transfer calls.")

D. *Data System Interfaces:* How your inbound ACD will work with your existing data-processing systems. Describe your data-processing environment in some detail. Your outbound ACD will need to work with it, since you'll probably be using the ACD for telemarketing, collections, or some similar purpose. In your description, include both the hardware (we do all our data-processing on a Diplodocus 3200M) and any relevant software subsystems. Ask your data-processing group to help with this section, especially if you use IBM mainframes. Depending on your application you'll need one or more of the following interfaces. Describe the ones that count as part of your RFP, so responders don't have to guess, and ask for a detailed description of the interface.
   1. *Order Processing:* If you're buying an outbound ACD to support a telemarketing operation, you need some sort of automated interface. You have three possibilities:
      a. *Batch:* At the end of each shift, your outbound ACD can bundle up all orders and dump them into your order processing system.
      b. *Real-time:* Every time one of your telemarketers finishes entering an order, the system adds that order record to your order processing system.
      c. *Integral data entry:* Let your telemarketers use the order-entry screens built into your order processing system (see above).

      Batch and on-line interfaces tend to be considerably less complicated and expensive than real-time links. On the other hand, you may promise overnight shipment. If you do, and if you choose not to buy a system based on open architecture, you have little choice.

2. *List Management:* We all know the rules on adding names to our house lists—we can add a name to our list if we get a response from that name. Make sure you can.

   You need the reverse link, too. Whether your application is telemarketing, collections, or whatever, you probably need to call out to a list you already have in-house. Make sure you can dump your existing lists into your outbound ACD. (See "List Conversion" for more on this subject.)

3. *Payroll:* Most telemarketing groups receive commissions. If you distribute commissions through your payroll system, this link is an obvious one. Make sure you test it *thoroughly.*

   Payoff Idea: Involve your data-processing group early in this process.

E. *Security:* Security isn't a big deal for an outbound ACD, but if you have a call center of any size you have some requirements. You'll want supervisor logins hidden behind passwords, because ACDs store employee performance information. Your administrator login must be secure, since it allows you to change the system configuration, and because you don't want anyone fiddling with commission records.

F. *Programming and Administration:* Don't ask for a lot of detail in your RFP, unless you're prepared to read system manuals from every vendor you solicit. Ask how you define new agents and customize data-entry screens, and how much training you'll need.

G. *Backup and Recovery:* Find out the system's provisions for tape backup. Especially, ask if you can fit a complete backup on a single tape, or whether you'll have to have someone present to switch tapes during the process. Also ask if you can backup the system while it's in use.

H. *Capacity and Expandability:* Describe your company, your call center(s), and the number of employees involved. Ask your vendor for recommendations on how much trunking and storage capacity (disk size) you'll need.

Also, ask how expandable the system is in terms of number of stations, trunks, and database records, and whether the vendor can integrate the operation of more than one system. You'll need this information if your initial size estimate was wrong or your organization grows.

Rules of Thumb Department: Typical call centers operate well with a 2:1 trunking ratio (two outside trunks for each call center agent).

2. Business Requirements
   A. *Warranty and Maintenance Terms:* Most warranties cover one year, but ask anyway. Maintenance contracts usually cost 10 to 12 percent.
   B. *Payment Terms and Conditions:* Typical terms are a third up front, a third on installation, and a third on acceptance. Ask.

Evaluating Responses: Review chapter 2 for tips on evaluating responses to RFPs. One important point: Outbound ACDs are a newer technology than the others you have dealt with. You may have vendors who will have important features "...in their next release of software." The industry has a sarcastic term for this: *Vaporware.* Don't rely on any features that haven't seen at least six months of commercial use.

INSTALLATION

You'll find yourself taking on more project management responsibilities than in most voice installations. Why? You have to coordinate your vendor's activities with those of your internal data-processing group.

Payoff Idea: For good and valid reasons, most data-processing organizations become obsessive regarding project management. Don't let them impose their project management paradigms on your project. They won't fit well. Data-processing organizations use methods appropriate for big, complicated projects where requirements change during system design, and lots of nasty little details have to come together.

Your role as project manager will be establishing a timetable, and asking everyone to stick to it. Here is a sample task list for installing a turnkey system. Modify it as you need to.

Absolutely Critical: Put dates on every step, and make sure *everyone* buys into the timetable.

Task	**Players**	**Description**
Prepare Location	Building Manager	Build or modify the room in which you'll install equipment. If you need raised flooring, power, or whatever, put it all together. Your building manager is used to these projects, and can figure out the details.
Install Equipment	Vendor	Get the equipment on site, install it, burn it in, and make sure it works.
Customize System	Vendor, Users	Customize data-entry screens, reports, commission calculations.
Establish Data Link	Vendor, Data Processing	However you plan to pass information between your own computer system and the outbound ACD, establish the link.
Test Transactions	Everyone	Run sample transactions through the outbound ACD and data link, to make sure everything works the way it's supposed to.
Train Call Center Agents	Vendor, Users	Make sure everyone knows how to use the new system.
Go Live	Users	With a long /. Start using the system.
Tune System	Everyone	For the first two months, hold weekly status meetings to discover bugs and fix them, and to review potential design improvements.
Go back to your job	You	Take a break. You've earned it.

A Few Tips

1. *Decide how to handle premature connections:* You have one important design decision. Since outbound ACDs use predictive dialing (the reason you bought the system, remember?) they'll sometimes make a connection before you have any agents ready to handle them. You have these options:
   - Hang up immediately—it may be rude, but business is business, and they'll never know you were the culprit.
   - Tell them what's going on—play a brief message such as, "Please stand by for an important message from the XYZ Corporation." This approach isn't fabulous, but it's honorable and inoffensive.
   - Dead air—don't say anything at all, and hope the wait is short enough it sounds like a caller catching his breath at the wrong moment. Opinion: don't do this.
   - Slander your competitor—play a long, tedious, phony sales pitch for your major competitor. Do your best to offend your audience. Your competitor could sue your brains out (although if you use this approach, you have no brains).

   Choose between the first two.
2. *Don't saturate your market:* An outbound ACD lets you make a *lot* of calls. If you run a telemarketing operation selling to a local market, you will run out of prospective customers long before your sales agents run out of breath. Do some arithmetic based on average call length, number of prospective customers, and how often you want to call each customer (the cycle time). Then staff your telemarketing center with the right number of agents.
   A. Customers per Agent per Day = Calling Hours/Call Length
   B. Agent Days for Total Market = Total Market/Customers per Agent per Day
   C. Staffing = Agent Days for Total Market/Desired Cycle Time
3. *Make sure you hang up:* Some outbound ACDs do a poor job of hanging up the phone, especially if the person they're calling hangs up first. You may need a phone service option called *Remote Disconnect* which breaks the connection when either party hangs up.

   Bad News: Depending on your local carrier and the Public Utilities Commission, remote disconnect can cost you significant money.
4. *Get on the data-processing change list:* Data-processing

departments generally have checklists of what to look at when changing their systems. Make sure the outbound ACD goes on their checklist, because a change in (for example) the order processing system can affect the interface with your outbound ACD.

Managing an Outbound ACD

Many of the activities that make phone system management such a headache (moves, adds and changes in particular) don't have much impact in an outbound ACD. Out-calling operations tend to be stable. You may need to add an occasional workstation, or someone may decide to rearrange the furniture, and you need to manage cabling and inventory just as you do with a phone system, but that's about it in most organizations.

Your call center manager will have some work to take on, in the form of running periodic reports, but that isn't your responsibility (unless you're the call center manager, in which case you'll receive training in how to run reports).

You will have two headaches to worry about: Changes in how the call center does business (commission changes or whatever), and changes in the data-processing system.

Minor changes in call center operation, such as commission changes, should not cause you any difficulties if you understand the basic process of customizing the system. You'll change the commission parameters and run through some test transactions. Your training should include instructions on how to do this.

Treat a major change in call center operation as you treated the original installation. In other words, get your vendor back in to take care of it. Most likely, a major change will affect your data-processing operation along with the outbound ACD. Whether you have a turnkey system or one based on open architecture, you have a significant project on your hands.

If your data-processing group changes part of the system to which your outbound ACD has links, you stand a good chance of having to alter the interface. Unless the change is very minor, bring in your vendor.

Payoff Idea: Pay your vendor either a retainer or a maintenance fee to manage your system for you. Unless you skipped the last two paragraphs, you won't find this surprising. You won't have day-to-day interaction with the system, and most changes are a matter of programming more than they are a matter of administration.

Old-hat Payoff Idea: Make sure the call center manager backs up the system to tape every day, and (as usual) stores at least one current backup off-site.

CHAPTER TEN

Interactive Voice Response Systems

THINK ABOUT AUTOMATED TELLER MACHINES (ATMs) for just a minute. Despite all preconceived ideas of technophobia in the general population, ATMs have succeeded in taking over a huge percentage of all consumer banking transactions. Why?

Compared with human tellers, customers find ATMs more convenient. You can find them where you need them, you can use them twenty-three hours a day (they do need servicing and refilling), they complete your transaction quickly, and they have no moods.

Don't you wish you could provide something like an automated teller?

If your customer transactions run through a computer, you probably can. The technology, *Interactive Voice Response,* lets your customers use the touchtone pad of their telephone to control your computer, prompted by voiced instructions. (A word on terminology: The technology is "Interactive Voice Response." Call the boxes themselves "Voice Response Units." When you talk with people in the industry, they"ll use the acronyms IVR and VRU.)

Business Summary

Useful Life:	3 years
Cash Cost:	$1,000 to $1,500 per telephone line answered
Amortized Cost:	$1,000 per telephone line answered per year
Dedicated Staff:	None
Justification:	Handles significant percentage of calls to call center without human intervention. Allows services not possible if staffing were needed.

Cost Analysis (estimated)

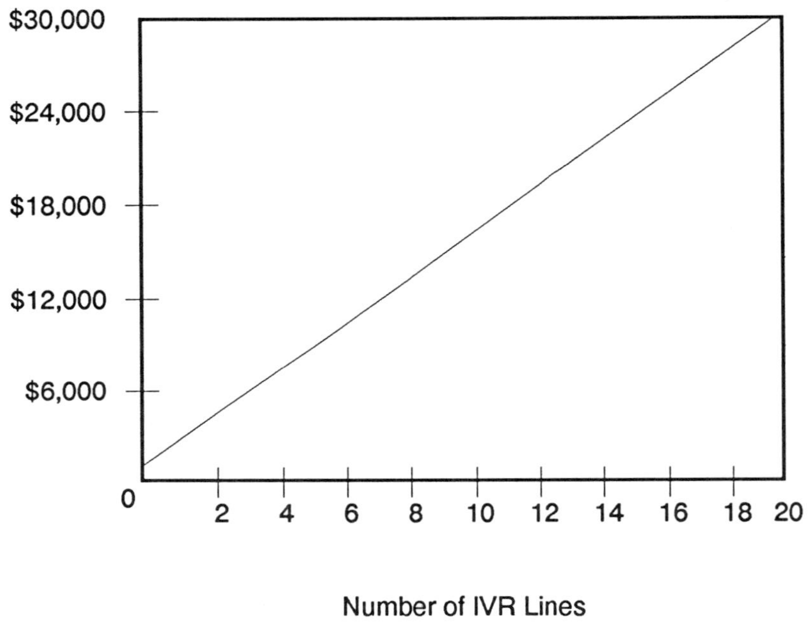

Examples of ways to use IVR:

- Transfer money from savings to checking.
- Stop your newspaper for the weekend.
- Order from a catalog.
- Get flight information.

And a huge list of others.

If you can get all transaction information from a touch-tone pad, you can use IVR. A typical application, in this case, customer order entry, might go like this:

1. Your voice response unit asks callers for their telephone numbers, using a digitized voice.
2. It uses the telephone number to bring up your customer's records from your database, and makes sure it has the right records.
3. The voice response unit presents a menu of possible transactions ("To order from our catalog, Press 1") and the customer responds. As with an automated attendant, one of the menu choices is to press 0 to talk to a live operator. (As an alternative, make this a special, heavily promoted service with its own telephone number.)
4. The system prompts your customer through the process of entering all information you need, and validates each. ("Enter the first item's catalog number, followed by the pound key." *Customer does so.* "You have ordered red aluminum widgets. Now enter the number you would like followed by the pound key. If you entered the wrong catalog number, enter zero." *Customer does so.*)
5. When finished, the VRU feeds your customer's transaction(s) into your order-entry system.
6. The system thanks your customer (you can rely on its manners) and hangs up.

Lots of companies have invested in interactive voice response, with two main motivations: Cutting costs, and offering new services.

Some applications (for example, order entry) represent cost-saving measures—doing the same things you're doing now, but at lower cost because you can have a machine do it. You can receive a hefty payback, but be careful you don't irritate your customers along the way. **Most Important:** Make sure your customer always has an escape to a live operator.

Interactive voice response can also give you a competitive advantage by letting you offer new services you couldn't offer before. For example, many brokerages and newspapers now offer stock quotation services. You dial in, key in a stock's trading code, and the system gives you an up to date quotation (either immediate or fifteen-minute delay). No company could afford to provide this as a consumer service using live operators. It would cost far too much. With interactive voice response, these companies can establish a new definition of "full-service provider," leaving their competitors to play catchup.

Some Technoid Stuff You Need to Know: IVR relies on five core technologies. You don't need to know a lot about them, but you do need to know what they are.

1. **Digitized Voice:** Words, phrases, sentences, and paragraphs spoken into the system and stored digitally for playback to callers.
2. **DTMF Signalling:** We've hit it before, but DTMF stands for "Dual-Tone MultiFrequency" and means touchtone except that Touchtone® is a registered trademark of AT&T. It means the ability to translate touchtone signals into numbers.
3. **Telephone System Interface:** Voice response units need to accept calls switched through your telephone system, and to transfer calls back to it (if a customer wants to speak to a live agent, for example). While more sophisticated ways exist, most systems act like a bank of analog telephones. Usually, the sales literature says something like this: "To your PBX our system appears as a cluster of standard 2500 sets." (A 2500 set is an old-fashioned telephone.)

[When you put together digitized voice, DTMF signalling, and a telephone system interface, you have an automated attendant. Here's what makes interactive voice response different.]

4. **Synthesized Speech:** Words, phrases, sentences, and paragraphs your system speaks by translating information in its database into sounds. The difference between digitized voice and synthesized speech: Digitized voice is the computer equivalent of a tape recorder—you can only get out what you put in. In synthesized speech, the computer stores the equivalent of written text, and reads it. (Some systems use a hybrid approach, in which people record speech fragments that the system assembles into speech. Think of Directory Assistance.)
5. **Computer Interface:** Most IVR systems include ways of getting information from an existing computer database, and possibly of updating that information.

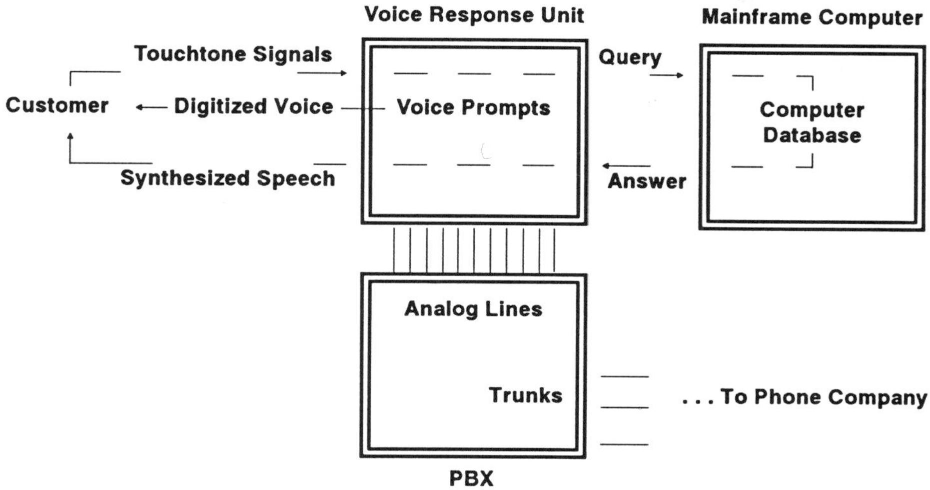

Figure 13.
Block Diagram, Interactive Voice Response

Figure 13 shows how interactive voice response combines these elements.

WHY A THREE-YEAR USEFUL LIFE

Every three years you'll have to either replace a voice response unit, or subject it to a major upgrade. Why? Some new developments in the industry, known collectively as open architecture, will transform interactive voice response into a new technology...*Custom Call Processing*. Custom call processing represents the marriage of voice processing and data processing. Chapter 11, page 169, provides some more information on this technology.

As with everything else you buy, you have to decide whether to issue a formal Request for Proposal, a less-formal Request for Information, or whether you should invite your vendors in for coffee, one at a time, and talk over what you need.

For IVR, you won't gain anything using the more formal methods. Invite your vendors in for coffee, give them a tour of your facility, and introduce them to your data-processing folks so they know what they have to deal with, and who to call for more information. You'll still want formal proposals, but interactive voice response systems all do the same sorts of things. Vendors differentiate their products by what mainframe systems they can interface with (and how), system capacity, skills needed to program them, and features like fault-tolerance that they'll tell you about whether you ask them or not.

If you choose not to write an RFP, you avoided little work. You've only dodged the actual writing. You still need to figure out your requirements and translate them into ways of evaluating the different systems you'll have to choose from. Here, as usual, you'll find some reasonable requirements. Don't take them as your own; use them as a place to begin.

Requirements

I. Technical Specifications

A. *Core Technology:* A controversial subject. Some experts believe you shouldn't care, so long as the system works. Others believe the subject is important because proprietary core technologies result in buggier systems and slower improvements.

Opinion: You should ask about four core technologies: The central processor, the operating system, the language used to create the system, and the scripting language used to program it. In the first three, you want indications the vendor hasn't reinvented the axle. [The wheel is a naturally occurring phenomenon, not an invention. The axle, which turned the wheel into a useful device, is an invention. (From my friend, Alan Bose.)] For the scripting language, you want assurance that the system encourages modification and tinkering.

For the central processor, commercially-available system boards are "best"; standard chips (from manufacturers such as Intel or Motorola) rank as acceptable. The operating system should be MS-DOS, Unix, OS/2, or perhaps another commercially-available operating system such as QNX.

Another reasonable approach for the central processor and operating system is to program on a widely used box, such as a Digital MicroVAX.

Nearly everyone uses the C programming language to create systems, but ask anyway.

Scripting languages fall into a family of languages called YAPL ("Yet Another Programming Language"). Vendors create their own, so expect no consistency. Ask for descriptions, and keep asking questions until you're comfortable. If your plans include contract support from the vendor, focus on how easy it is to make changes. If you plan to contract support from other sources, also focus on how well the vendor has documented the language, and the policy on software revisions.

If you plan to write your own scripts, go into a *lot* of detail in your questions. Ask about training, telephone support, and policies on new releases and revisions of software. Ask to see sample scripts and have the vendor walk you through them.

B. *Features:* Almost by definition, all IVR systems have the same basic features, so don't list them and ask, "Does your system do this?" You'll give the same impression as a customer at a car lot asking the salesman, "Does this car come with an engine and four tires?" (Well, at the car lot you may *need* to ask. . .)

   Ask another open-ended question instead: "What unusual features does your system provide that I'm not likely to find in competing systems?"

C. *Interfaces:* The subject providing the greatest differentiation among competing systems. Interfaces cover a lot of ground.

   1. *DNIS Handling:* DNIS stands for "Dialed Number Identification Service" and means you can route more than one dialed number to the same phone. Lets say you have more than one call center (sales and service) and you want to provide interactive voice response for both. A voice response unit that handles DNIS can tell who the caller wanted to reach, and handle the call properly without having to ask.

      (Don't bother. Voice response units don't cost that much. Buy two smaller ones instead, one for each call center. Once you get into DNIS, you have to assign different scripts to different ports, and your life becomes much more complicated than it has to be.)

2. *ANI Handling:* As mentioned elsewhere, ANI stands for "Automated Number Identification" and it means you know who called you. A voice response unit that handles ANI doesn't have to ask for a caller's phone number. It does have to confirm it, since not everyone calls from the telephone in your database (". . . I'm calling from work . . .").

   Ask your local and long-distance service providers when they can provide you with ANI, and with what forms of service.

   Favorite Application: A voice response unit in an order processing department. It takes ANI, looks up customer records in its database, and transfers calls from delinquent accounts directly to the Collections Department.

3. *Rotary Telephones:* A surprising number of callers, especially residential callers, still use rotary telephones. Most voice response units can't handle them. If this market segment is important to you, ask about pulse-to-DTMF converters. They exist, and they're pretty good. If at all possible, make your vendor responsible for selection and installation.

4. *Voice System Interfaces:* How your new system will hook up to your existing phone system. Make sure your vendors know what they're up against—the make, model, and age of your telephone system and inbound ACD (if you have one), number and type of central office lines or trunks, and number of telephones, modems, fax machines, and whatever else your system supports.

   Ask for *specifics* in terms of how the system transfers calls. Most likely, you'll want your voice response unit to transfer calls to your ACD when it can't handle a transaction itself. As mentioned earlier, most systems emulate analog telephones and transfer calls the same way anyone else does—through touchtone signals.

   Ask for a block diagram and a detailed description of how the VRU will hook up to your existing phone system. Beware of generalities. Statements like, ". . . all we need is an RS-232 port to hook up to . . ." betray a dangerous lack of insight. Similarly, the phrase "We've never worked with your exact make and model, but it shouldn't present any special difficulties," should cause the hairs on the back of your neck to stand up, unless the word "because" follows it (". . . because as far as your system is concerned, our system looks just like a rotary hunt group of 2500 sets that use standard touchtone signals to transfer calls.")

Don't be surprised if all block diagrams look just like Figure 13. It's the most straightforward way of handling the interface.

5. *Data System Interfaces:* The trickiest subject. Make sure your vendors know what hardware and operating system you use, and give them the name and phone number of your data-processing manager, or whatever analyst the manager delegates this to.

Mainframe computers, especially those from IBM, have so many different ways of hooking together that it's easy for your vendors to make incorrect assumptions regarding what they'll find to attach to.

The most common way of linking to mainframe computers is to emulate a bank of standard computer terminals. In other words, you or (better) your vendor will have to teach your voice response unit what the relevant mainframe data-entry screens look like, how to read information from them, and where to enter new transaction information. Your VR Unit reads screen information *from*, and feeds keystrokes *to* what your computer thinks is a standard data-entry screen.

Key Question: "Will I have to modify my mainframe applications at all to make this work?" The answer should be "No."

More Seldom-Asked Key Question: "What kinds of change to my mainframe applications will require me to also change the programming of my voice response unit?" The answer is "Any change to the data-entry screens you're using." With the ideal hookup, which does *not* work by emulating a terminal, the answer would be, "Only changes to the structure of your database."

D. *Security:* You need to secure voice response units. The process is neither onerous nor paranoid, and it takes little effort. All you have to do is:
   1. Physically restrict access to the system—if you have a computer room, put it in there (it's small); otherwise some other room that is normally locked will do. (Avoid janitorial closets. These usually have sinks in them, and sinks mean water.)
   2. Hide administrative functions behind a password (you'll find this subject covered in more detail in the section on Local Area Networks (LANs) in the companion volume to this book).
   3. If your vendor needs a modem for remote diagnosis, maintenance, and system reprogramming, make sure *all* logins it has access to are hidden behind randomly generated passwords (if you work for General Motors, "acqbd" is a much better password than "chevy").
   4. Even better, insist that all modems use "dial-back" security (also described in *Communications Systems for Every Business*).

   A poorly secured VRU creates a hole in your mainframe computer security. It also gives any practical joker with a modem and leisure time an opportunity to make you look foolish in front of your customers. You have a hole in your mainframe security because the voice response unit must have a legitimate login sequence stored in a file somewhere. Hackers can make you look goofy because you've given them a modem to call, which they can use to reprogram your system.

E. *Programming:* Ask if ordinary mortals can easily learn to program the system (if you are prepared to take on that task; otherwise you don't much care about this subject).

F. *Backup and Recovery:* Find out the system's provisions for tape backup. Especially, ask if you can fit a complete backup on a single tape, or whether you'll have to have someone present to switch tapes during the process. Also ask if you can backup the system while it's in use.

G. *Capacity and Expandability:* If you want to use interactive voice response to automate part of an existing call center, you can easily figure out how big a system you'll need. You need to know (1) how many agents you keep busy during peak calling hours; and (2) what success rate you expect (how many callers you think will use your voice response unit instead of dropping out for a live agent).

 Multiply the success rate by your number of agents. That's the number of dial-in ports you need. (The number of ports is the same as the number of simultaneous calls.)

 This arithmetic (if you're a consultant, call it a "computation") works so long as callers won't need more time to deal with your voice response unit than they do with a live operator.

 Payoff Idea: Make *absolutely sure* your VRU won't take more of your customers' time than live operators. **The whole point is to IMPROVE the service you provide.** Saving money by annoying your customers is a poor idea.

 Ask how many ports a vendor can put in a system, and the increments. You care about the former because that gives you the safety margin. (If you buy an eight-port system and your vendor can handle thirty-two ports, you have a lot of headroom.) You care about the latter because it tells you how easy it is to expand. (If ports come two per card and you can put eight cards in a system that limits you to sixteen ports per system. You can hook up multiple systems, but if you start with a sixteen port system your next expansion will be expensive, making you buy a second system unit.)

H. *Training:* Ask about training if you plan to maintain the system yourself. If you don't want to learn how to program interactive voice response scripts, don't bother asking.

 Unless you expect to make frequent changes to

your scripts (in which case you'll need more help than this book can give you) you're better off paying your vendor, who can do it faster. You'll have that much less recordkeeping to worry about.

Payoff Idea: Ask how much each vendor charges to make different types of script changes. Examples: (1) Changes to wording only; (2) Changes to the sequence of events; (3) Changes to mainframe data-entry screens.

2. Business Specifications
   A. *Warranty and Maintenance Terms:* Most warranties cover one year; maintenance contracts usually cost 10 to 12 percent.
   B. *Payment Terms and Conditions:* Typical terms are a third up front, a third on installation, and a third on acceptance. Ask.

Evaluation: Most of the above are pass/fail criteria. You're checking to make sure each vendor does a decent job.

Once you've finished that initial screening, focus on voice and data interfaces and system capacity, and charges for script changes. You'll make your decision based on answers to those questions.

If you plan to program the system yourself, make sure you select an MS-DOS system. Systems based on MS-DOS may not be the most powerful, but here's a trade secret: *That doesn't matter.* A Chevy Caprice may not have the horsepower or torque of a Mack truck, but you use it to commute, not to haul freight. Similarly, MS-DOS systems may lack the horsepower found in OS/2 or Unix, but so long as they're powerful enough for you (most likely they are, since you can configure some of them with as many as forty-eight ports) *who cares?*

With a system based on MS-DOS you're using off-the-shelf PC technology, which will make troubleshooting and debugging much easier.

You have one responsibility in evaluating vendors that doesn't fit into any neat category: Making sure the vendor you pick won't need to learn on the job. Technical salespeople too often lack technical understanding of their products. Before you buy, talk directly with someone who understands the product well enough to answer every question you can think of. Ultimately, if an installation goes haywire, you failed to ask an important question.

Installation

As with many of the enhanced call processing systems (autoattendant, outbound ACD) you will have to act as project manager when installing a voice response unit. You won't have to actually *do* anything, but you can't contract out responsibility for assigning roles and responsibilities, nor can you contract out responsibility for adhering to schedules.

You can't, because you'll have contractors waiting on inhouse personnel for many of their activities.

Much of what was true for installing an outbound ACD also fits an interactive voice response installation:

Payoff Idea: Most data-processing organizations become obsessive regarding project management. They have to. You don't, so don't let them impose their project management paradigms on your project. Data-processing organizations use methods appropriate for big, complicated projects where requirements change during system design, and lots of nasty little details have to come together.

Your role as project manager will be establishing a timetable, and asking everyone to stick to it. Here is a sample task list. Modify it as you need to. Make as much happen in parallel as you can.

Absolutely Critical: Put dates on every step, and make sure *everyone* agrees the timetable is reasonable.

Task	Players	Description
Prepare Location	Building Manager	Figure out where you'll put the thing. It won't be much bigger than a large personal computer, but you'll probably want an UPS to go with it. Figure you'll need nine square feet (a three-foot square if your math is weak). You'll also want a lockable door.
Install Equipment	Vendor	Get the equipment on site, install it, burn it in, and make sure it works.

Develop Menus and Scripts	Users	Figure out exactly what you'll want the system to say at each step of the proceedings. Refer to the guidelines in chapter 6, Automated Attendants.
Program System	Vendor, Users	Implement menus and scripts. Your vendor will program; your users will record scripts. **Payoff Idea:** Use a professional, and interview several to find one with a voice you like.
Establish Data Link	Vendor, Data Processing	However you plan to pass information between your own computer system and the voice response unit, establish the link.
Test Transactions	Everyone	Run sample transactions through the VRU and data link, to make sure everything works the way it's supposed to.
Go Live	Users	With a long /. Start using the system.
Tune System	Everyone	For the first two months, hold weekly status meetings to discover bugs and fix them, and to review potential design improvements.
Go back to your job.	You	Take a break. You've earned it.

A Few Tips

1. *Provide an escape, and tell your callers about it.* Some callers just won't like talking to a machine. Unless you're using IVR to provide a new information service and you simply can't afford to direct any labor toward it, make sure callers can press 0 to reach an operator.

   Do you make the escape first or last in your presentation? Proponents of the former claim it emphasizes customer service. Proponents of the latter claim you get the best results (translation: highest utilization) by presenting the escape option last.

   If you expect callers to use interactive voice response often, present the escape last. They'll get the best service from your VRU. If you expect a lot of one-time callers, you may be better off presenting the escape first (but then, if they're one-time callers, why not maximize the return on your investment?)

Urgent Recommendation: Whatever you do, use 0 for the escape. It has reached the status of *de facto* standard, so frequent users of these systems can anticipate it.

2. *Make sure your system offers more convenience than live operators.* Dial into your call center and time a transaction, including time spent on ACD hold. Now role play the same transaction using IVR. Did it take substantially more time? If so, edit your scripts to cut as many superfluous words as possible. Remember, every syllable takes precious time.

   One Saving Grace: Caller time flows more slowly on hold. In other words, a transaction that takes five minutes using IVR seems shorter than the same transaction that takes three minutes with a live operator, if the caller waited on hold for two minutes before reaching the live operator.

3. *Review all guidelines for automated attendants, and apply them to your voice response unit.* Guidelines for voice response menus and scripts match those for autoattendants because. . . well, it's obvious, isn't it?

4. *If part of your process involves looking up caller records in your database, be careful how you validate caller identification.* For many applications you may need nothing more than, "Are you Irving Glotz of 123 Main Street? Press 'Y' for 'Yes' or 'N' for 'No'." This approach does nothing to discourage practical jokers, nor does it help those callers who tend to filter conversations. (This class of caller would hear the above dialog as, "Blah blah blah blah blah blah blah? Press 'Y' for 'Yes' or 'N' for 'No'." To which they automatically press Y, muttering "Get on with it" under their breath.)

   A Better Approach: Ask callers to press the first two digits of their street address instead.

5. *Get on the data-processing change list:* Data-processing departments generally have checklists of what to look at when changing their systems. Make sure your voice response unit goes on their checklist, because a change in (for example) the order processing system can affect its interface.

Managing a Voice Response Unit

Managing a VRU is similar to managing an outbound ACD, and for a very good reason: They have similar architectures and interfaces to your other systems.

Very simply, you don't have much to do unless things break, and if you're smart you still won't have much to do.

In many situations, you are best off contracting with your vendor for support and maintenance, unless you have deliberately selected a system that you can maintain yourself.

Payoff Idea: Buy a second system with one port in it. Use it for new programming and testing so you're not developing software on a production system. As a fringe benefit, you have a spare for those occasions when the first system goes belly-up. Keep a copy of your production software on the development system's hard disk. If you have to use the spare, just swap in the boards from the first system and turn on the power. (It's never really *that* simple, but that's the idea.)

In addition to your ever-popular task of keeping track of cable (don't think of it as a chore; think of it as job security) and the other standard stuff, you have the same responsibilities for a voice response unit that you have for an outbound ACD. You have to respond to operating changes, changes in how your organization wants to handle calls, and changes to your data-processing systems.

Treat operating changes as you did the original installation, only easier. You or your vendor has to reprogram the system to behave as desired.

For changes to the data-processing system, bring in your vendor. You don't want to fool with that stuff unless you're *seriously* underworked.

Besides, significant changes to the data-processing system probably signify serious operating changes as well.

The Usual Payoff Idea: Make sure you backup the system to tape every day, and (as usual) store at least one current backup off site.

One Last Payoff Idea: Don't fall in love with the technology. One more time—these are your customers we're talking about. At each stage of planning, think, "Would I enjoy calling into this system, or would I lose patience and escape to a live operator?" Your most important skill is *customer empathy,* not use of technology. Practice it every day.

Interactive voice response can provide huge payoffs for most organizations. Deciding to exploit this technology may be as close to "no-brain-required" as most people will ever face. When a $50,000 box can handle as many calls as sixteen paid operators, you have a hard time avoiding the investment.

CHAPTER ELEVEN

The Future of Voice Communications

PREDICTING THE FUTURE presents more of a challenge than other forms of prediction... predicting the past or present, for example. Still, a few trends seem clear. Take a look, then, at ten predictions from the author's crystal ball:

1. **Price Collapse:** The price you pay for key systems, PBXs, and adjunct processors such as ACDs, autoattendants and voice mail has fallen in recent years, and will continue to drop. The reductions won't be dramatic, but contrasted with inflation they will make this technology increasingly accessible to the average business.
2. **Confusion, in the form of Lawyers, Reigns:** Nine years ago, Judge Greene broke up AT&T into Snow White and the Seven Dwarves (that is, AT&T and the seven local Regional Bell Operating Companies). Since that time, despite the best efforts of sales and operating staff, the process of ordering and receiving service has become at best a peculiar experience. The legal profession created the business units and the rules under which they operate. You could easily predict the result: Companies that operate fairly and openly, but with no notion of operational efficiency or competitiveness.

3. **Confusion, in the form of Weird Sales Offerings, Reigns:** Local and long-distance services will offer an array of services whose range and complexity will continue to increase. This will continue for awhile, but an optimist might hope that some industry leader will find a way of packaging that complexity into an understandable set of features, and bring order out of chaos. Prediction: Chaos. But we can hope.

4. **ISDN:** It *Still* Dials Numbers. ISDN will stagger into the marketplace as a competitive offering, once the phone companies realize their customers won't pay big bucks for its so-called advanced features. ISDN benefits the *telephone companies,* not their customers. Once pricing becomes competitive, customers will cheerfully hop on the bandwagon. It does provide some useful features, and a clean architecture, which companies can use to lower their operating costs and increase the level of service they provide to their customers.

5. **Price Reductions, Round Two:** The price you pay for long-distance service will continue to drop, fueled by an increase in nationwide capacity. These reductions will fall directly to your profit line, since few companies will loosen their long-distance dialing policies. Another consequence:

6. **Fax Machines:** Offices won't see an end to paper for a long time, but the fax machine has had significant impact on express delivery services. Combination fax and answering machines will become the next hot consumer item. They will automatically detect the *CNG* tone produced by most fax machines and use it to print fax calls while acting as an answering machine for all others. Fax in the home will give consumer-oriented companies a whole new vehicle for sales and service.

7. **Home Businesses and Telecommuting:** These have become real issues in recent years. Fueled in large part by the low cost of computers and telecommunications, significant numbers of adults will find themselves working in their homes, rather than commuting to work.

   Home businesses can now produce products indistinguishable from those produced by companies hundreds of times larger. Similarly, employees can produce much of their work from home, without outside customers knowing the difference.

8. **Wraparound:** Not a telecommunications trend at all, wraparound refers to the business practice of enhancing products with ancillary services that turn commodity items

into differentiated offerings. A single example: Northgate Computers sells PC clones. Unlike many of their competitors (they have hundreds of competitors), Northgate offers a guaranteed upgrade policy and twenty-four-hour telephone support.

Telecommunications will provide a vital component of most wraparounds. For example—could Northgate offer twenty-four-hour telephone support without telephones?

9. **Speech Recognition:** Within two years, a limited, but very useful form of speech recognition called *Word Picking* will radically change the way we interact with autoattendants. Instead of a tedious dialog in which the machine prompts callers to press buttons ("For Customer Service, Press 1. . ."), the machine will ask, "To which department may I direct your call?" Out of whatever response the customer happens to provide, ("I'd like the Customer Service department, please,") the system looks for any of a list of a hundred or so words or phrases.

   The systems you'll see on the market sooner than you think will even understand southern and New England dialects. Californian ("Like, this is awesome. . .like, I'd like, ya know, the guys who help you when you're not sure. . .") may continue to elude this technology. And the rest of us.

10. **Custom Call Processing:** The blockbuster technology of the next decade. The logical extension to interactive voice response, *Computer-Controlled Telephony* will merge voice processing with data processing. In typical scenarios, VRUs answer calls and interact with callers to identify them. They then look up information about the caller from the company database, and route or process the call, all without any human intervention. The distinctions between automated attendants, ACDs, and VRUs will disappear. Typical scenarios:
    A. *Tandem Voice/Data Switching:* When an ACD routes your call to an agent, your customer records have already appeared on the agent's computer screen.
    B. *Sales Group Call Routing:* The ACD that handles your call transfers it to the sales group responsible for your geographical region.
    C. *Collections:* Using automated number identification (see chapters 3 and 5), your voice processing system will identify everyone calling into your sales center. The system will automatically route those with past due accounts to your Collections Department, even though they called to make another purchase. *You can do this right now! Some companies do!*

These prognostications have the reliability you'd expect. Don't invest in the stock market based on this advice, and if you do, don't complain to the author, publisher, or their respective legal representatives. For investing, you're on your own.

Do use these predictions as you make strategic business plans. The author didn't invent any of them—they're commonplace in the trade press—so you can be sure your competitors will invest in some of these technologies. Some of them have the disadvantage of being much bigger than you, so you should be able to beat them at their own game.

Size has its advantages, but telecommunications can be the great equalizer. Use it.

APPENDIX A

A Computerized Telecommunications Management Database

If you want to set up a computerized telecommunications database, use one of the popular Relational Database Management Systems (RDBMS) for personal computers. Use a full-featured package rather than an easier-to-use records manager—this is not a trivial undertaking. A key feature to look for: The ability to display and edit information from more than one file in a single data entry screen.

If you have used dBase, Paradox, or any similar package what follows should look familiar. **Field Names** are restricted to eight characters, typical for an RDBMS. **Field Types** are restricted to *Alphanumeric* (any displayable character), *Numeric* (numbers that can be used in computations), *Date* (dates stored in a format that allows computations to be performed on them), *Logical* (Yes/No), or *TEXT* (a string of characters of indefinite length). The RDBMS you select may provide a longer field name or a richer set of field types. If so, take advantage as appropriate.

In file descriptions, alphanumeric fields are described as A*n*, where *n* is the number of characters in the string. For example, an A30 field is a field that can contain a string of up to thirty characters.

The other field types are described as N(umeric), D(ate), and L(ogical).

To keep track of your phone system, you'll need a multi-file database structure. Also, we're planning ahead: We can reuse the same database to keep track of many other things as well.

Payoff Idea: Many call detail reporting systems have evolved into complete telephone management systems. If you're planning to purchase one (you'll find dozens on the market), make sure you select one that covers what you need. Use this appendix as a guide, but not as a specification. Both the details of database design and your actual requirements may vary.

Keeping Track of System Information

The first group of files you create defines the system itself, and its components.

System Master File—File Name = SYSTEMS

Field Name	Type	Description
DEVICEID	A10	An identifier for the device you are managing (you will use this file for more than just your phone system). For your phone system, call it something like "TEL—1." If you grow large enough to manage more than one system you can call the new ones "TEL—2" and so forth.
DEVTYPE	A10	A classification field, defining the type of device you're defining. For a PBX, enter "PBX." For a key system, enter "KSU."
USERID	A10	Principal user of the device. For a phone system, use "TELECOM1." This field cross-references to the User database, described later in this section.
OUTCABLE	A10	If the device being tracked is a concentrator or multiplexor, it will have many cables as input and only a single cable as output. Leave this field blank for a phone system. (See section on Cable Management).
CABTYPE	A5	Type of cable (because you'll use this file for more than just telephone systems).

Primary Key[1]: DEVICEID

Other Keys: USERID, OUTCABLE

Other keys are fields which may not be unique, but are useful in identifying a record. Specifically, these are fields you may wish to use to rapidly locate a record.

Ports Master File—File Name = PORTS

Field Name	Type	Description
DEVICEID	A10	Matches the value of DEVICEID in the System Master File.
CABINET	A5	Cabinet number as designated by your vendor at installation. (Alphanumeric in case your vendor's practice is to count them as A, B, C, etc.)
SHELF	A5	Shelf number, counted from the top of the cabinet.
CARD	A5	Card number, counted from the left of the shelf.
PORT	A5	Port number, counted left-to-right, top-to-bottom as appropriate for the specific card.
CARDTYPE	A5	Type of card.
EXTNSION	A4	4-digit extension of the telephone to which the port is connected.
INCABLE	A10	Cable that comes into the port (see section on Cable Management).
CATYPIN	A5	Type of input cable. Enter DTUTP for "Digital Telephone Unshielded Twisted Pair" for digital-grade telephone wire, ATUTP for "Analog Telephone Unshielded Twisted Pair," and so forth. Also use this field (with INCABLE) to keep track of the trunks coming into your PBX (or the lines coming into your key system). Enter "1FB" for standard business lines, "DID" for Direct Inward Dial trunks, "DSS" for Digital Switched Service, and so on.
OUTCABLE	A10	Cable that comes out of the port (see section on Cable Management)
CATYPOUT	A5	Type of output cable. Enter DTUTP for "Digital Telephone Unshielded Twisted Pair" for digital-grade telephone wire, ATUTP for "Analog Telephone Unshielded Twisted Pair," and so forth.
USENOTE	TEXT	Usage notes. Any special information about a particular port beyond the above entry fields.

1. In an RDBMS, the *Primary Key* is the information with which you can uniquely identify each record. Here, the DEVICEID must be unique. Sometimes, a combination of several fields must be used to establish a unique primary key. When this is the case, those fields will be listed, connected by " + " signs, for example:
DEVICEID + CABINET + SHELF + CARD + PORT for the PORTS file.

Primary Key: DEVICEID + LOCATION + CABINET + SHELF + CARD + PORT

Note the absence of *User Name, Phone Type,* and *Location* from the database. In a computerized database, you'll be able to pick these up from a different file, linked by the phone number (described below under "User Management.")

In addition to this primary file, you'll want to set up validation files for *Card Types* and *Cable Types.* The CARDTYPE file includes a record for each valid type of card. The CABLTYPE file includes a record for each valid type of cable, because you can use this same database for Local Area Network management as well. The validation files have the layouts:

Device Type Validation File—File Name = DEVTYPE

Field Name	Type	Description
TYPE	A5	Type of system. Matches field in SYSTEMS file.
DESC	A30	Full description of the system type.

Card Type Validation File—File Name = CARDTYPE

Field Name	Type	Description
TYPE	A5	Type of card. Matches field in PORTS file.
DESC	A30	Full description of the card type.

Primary Key: TYPE

Cable Type Validation File—File Name = CABLTYPE

Field Name	Type	Description
CABTYPE	A5	Type of cable. Matches fields in SYSTEMS and PORTS files.
DESC	A30	Full description of the cable type.

Primary Key: CABTYPE

A database schematic diagram shows the relationships among the various files. This book follows standard conventions in using arrows to reflect "One to Many" relationships (also called 1:M relationships). A 1:M relationship simply means that for a single record in one file you may find many matching records in a related file. In the current example, a sin-

Database Schematic Diagram

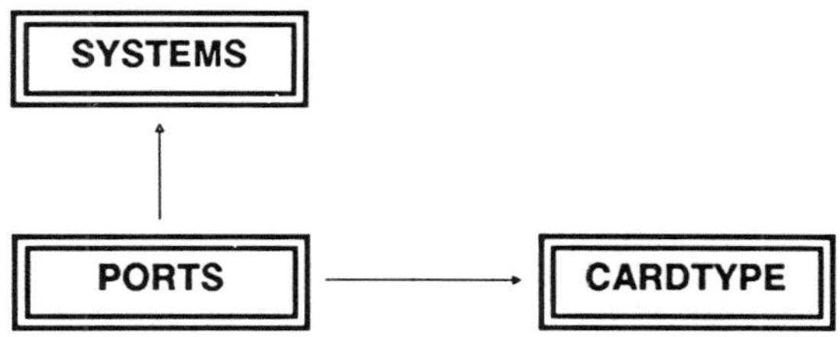

gle DEVICEID in the SYSTEMS file may match many different records in the PORTS file.

The arrow points to the "One" side of the 1:M pair.

Figure 14 shows what a typical data-entry screen might look like. In this and all following data-entry screens, data-entry fields are displayed as ████████.

One thing to note in Figure 14: The screen has room for the user name. This is a validation field, not a data-entry field. When you type the extension, you want the database to automatically look up the user's name (file described below) and present it.

Keeping Track of Trunks and Lines

You won't find great benefit from keeping track of outside trunks and lines on your computer. On the other hand, it won't cost you very much to set up your database files for it, either. In fact, it won't cost you anything, because you already have the data structures in place. Look in the PORTS master file and you'll find fields for the trunk or line ID (INCABLE), type (CABTYPIN), and usage notes (USENOTE). Instead of defining a special field for the *Carrier,* define each outside carrier as an additional device on your network. Enter OUTCABLE and CABTYPOUT data to match your inbound entries. *Voilà!* You're done.

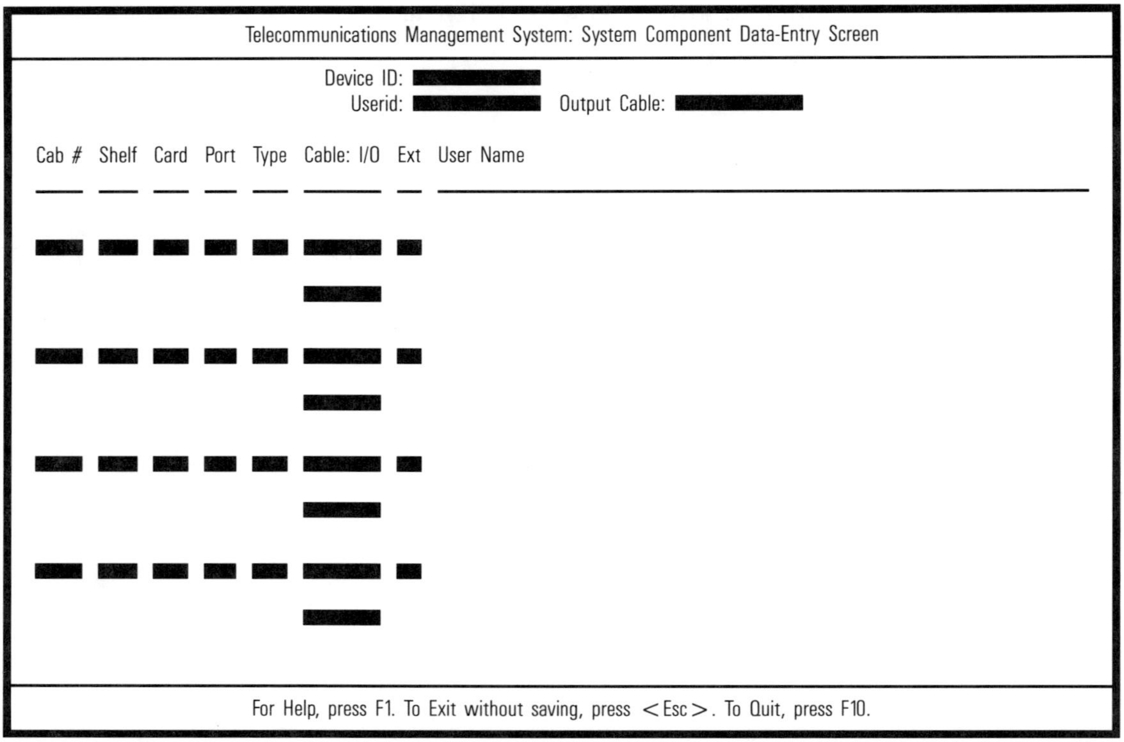

Figure 14.
Sample Phone System Data-Entry Screen

Keeping Track of Extensions

There are two ways of laying out the files that describe users and extensions. The next table is a description for the extensions file that works for key systems where all phones have the same number of appearances.

User Master File—File Name = USERS

Field Name	Type	Description
USERID	A10	A unique identifier for each employee. It's annoying to have to construct these, but even a small organization may have several John Andersons or Betty Smiths. Three possibilities: employee number from your payroll system; social security number; or a combination of last name, first name, and middle initial (John C. Anderson might be ANDERJC).

LNAME	A15	User's last name
FNAME	A10	User's first name
LOCATION	A30	Room number or equivalent location description
PHONTYPE	A10	Type of telephone
QUANTITY	N	Number of telephones (defaults to 1. See the section on Inventory for usage details.)
CABLE	A10	Number assigned to the cable running from the phone jack to the wiring closet
CLOSET	A30	Wiring closet to which the cable runs, identified by room number or whatever is appropriate
EXT1	A4	First appearance on telephone (primary extension)
EXT2	A4	Second appearance on telephone
.	.	.
.	.	.
.	.	.
EXTn	A4	Last (nth) appearance on telephone
COVER	A4	Extension of coverage path
GROUP	A10	Name assigned to call pickup group
CLASS	A10	Class of service as defined in the KSU (if the KSU uses class-of-service definitions)

Primary Key: USERID

Other Key: LNAME, EXT1

Figure 15 shows what a typical data-entry screen might look like.

The file structure for a PBX is more complicated. Since most phones will be single-line sets, reserving space for multiple extensions is poor practice. Instead, we'll set up a multifile database structure.

User Master File—File Name = USERS

Field Name	Type	Description
USERID	A10	Unique identifier for each user
LNAME	A15	User's last name

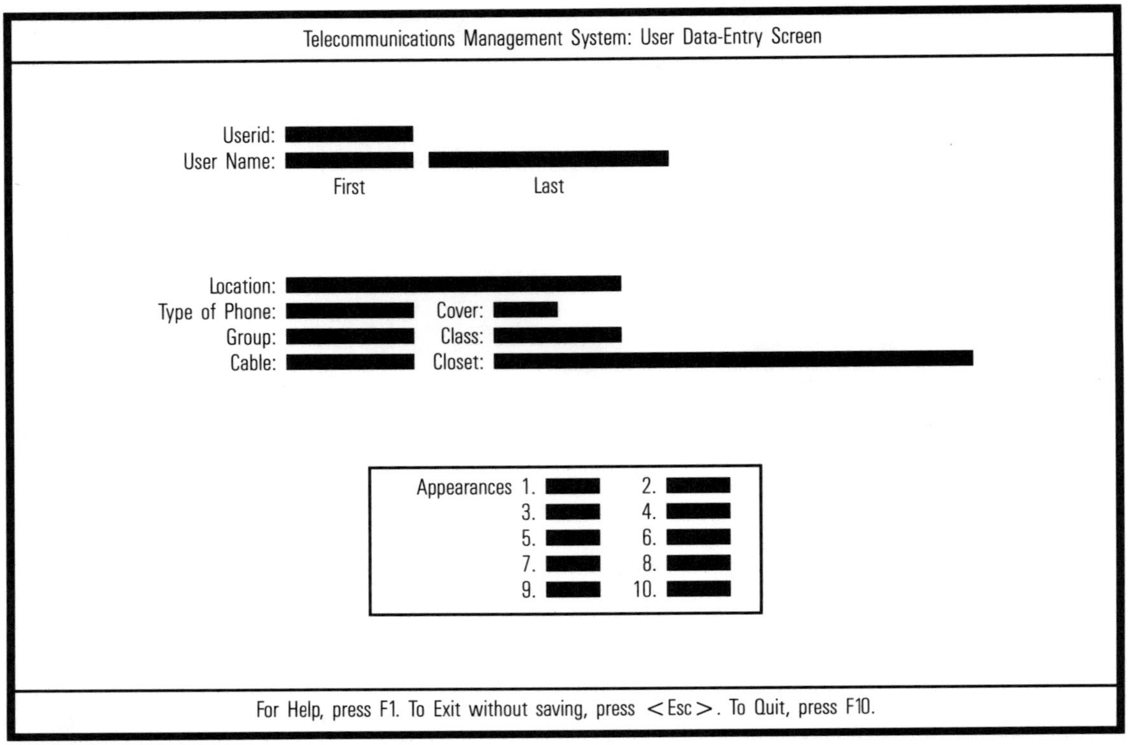

Figure 15.
Sample User Data-Entry Screen

Field Name	Type	Description
FNAME	A10	User's first name
LOCATION	A30	Room number or equivalent location description
DEPT	A10	Department the user works in
PHONTYPE	A10	Type of telephone
CABLE	A10	Number assigned to the cable running from the phone jack to the wiring closet
CLOSET	A30	Wiring closet to which the cable runs, identified by room number or whatever is appropriate
QUANTITY	N	Number of telephones (defaults to 1. See the section on Inventory for usage details.)
COVER	A4	Extension of coverage path
GROUP	A10	Name assigned to call pickup group

CLASS	A10	Class of service as defined in the KSU (if the KSU uses class-of-service definitions)
Primary Key:	USERID	
Other Keys:	LNAME, PHONTYPE CABLE	

Extensions File—File Name = EXTENSNS

Field Name	Type	Description
USERID	A10	Unique identifier for each user
EXTNO	N	Sequence number of extension—what order it appears on the phone
EXTENSN	A4	Extension assigned to that appearance on the phone
LNAME	A15	User's last name
FNAME	A10	User's first name
LOCATION	A30	Room number or equivalent location description
Primary Key:	USERID + EXTNO + EXTENSN	
Other Keys:	EXTENSN, LNAME	

You should have a validation file for *Departments.* The simple version is shown in the Departments File chart.

Departments File—File Name = DEPTS

Field Name	Type	Description
DEPT	A10	Department code
DEPTNAME	A30	Full department name
LNAME	A15	Last name of your liaison in the department (from the installation phase)
FNAME	A10	First name of your departmental liaison
Primary Key:	DEPT	

Figure 16 is a data-entry screen that combines the files. The *Appearances* box scrolls up and down to accommodate as many extensions as necessary.

In addition to these files, you should establish several validation files, to make sure entries in the PHONTYPE,

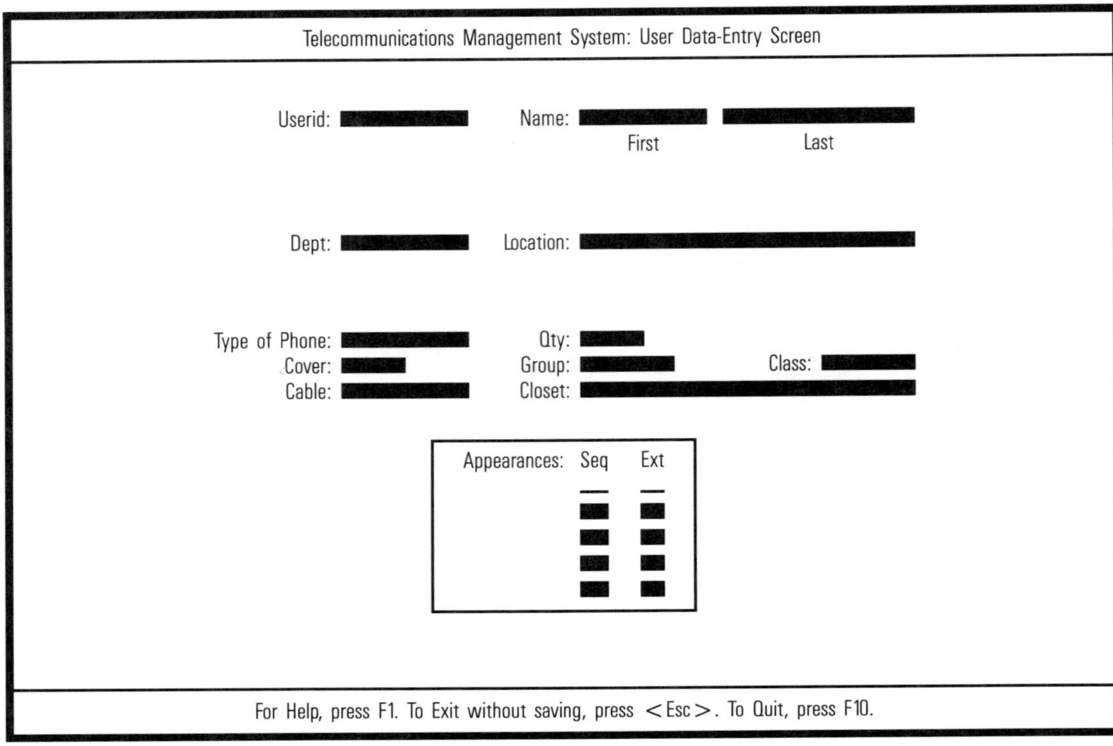

Figure 16.
Sample User Data-Entry Screen

GROUP, and CLASS fields use standardized notations and spellings. (See Telephone Type Validation File chart.)

Telephone Type Validation File—File Name = PHONTYPE

Field Name	Type	Description
PHONTYPE	A10	Type of telephone. Matches field in User Master File.
DESC	A30	Description of the phone model
APPEARS	N	Number of appearances (how many lines) on the phone
SPEAKER	L	Is it a speakerphone?
DISPLAY	L	Is it a displayphone?
OTHER	A255	List any other significant features of this model.

Primary Key: PHONTYPE

Group Validation File—File Name = GROUPS

Field Name	Type	Description
GROUP	A10	Name of group. Matches field in User Master File.
DESC	A30	Description of group (department name or equivalent)

Primary Key: GROUP

Class Validation File—File Name = CLASSES

Field Name	Type	Description
CLASS	A10	Class of service. Matches field in User Master File.
DESC	A30	Description of class of service (if appropriate)
ATTRIB1	A10	First attribute of the class of service (as defined in the PBX administrative system)
ATTRIB2	A10	Second attribute
.	.	.
.	.	.
.	.	.
ATTRIBn	A10	Last attribute

Primary Key: CLASS

Database Schematic Diagram

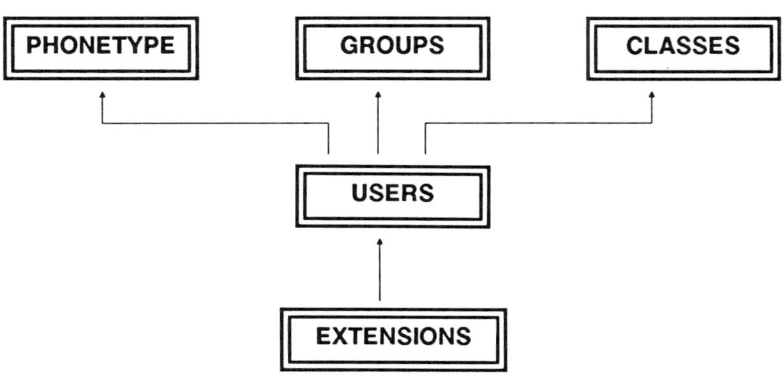

The class of service record is where most user parameters are stored. For some systems you may prefer to eliminate the CLASSES file and instead put ATTRIBUTE fields directly in the User Master Record. Use the style that will work best with the administrative software in your PBX.

Obviously, you won't want to call the attribute fields ATTRIB1, ATTRIB2, and so forth. Give them explanatory names.

Keeping Track of your Telephone Inventory

You already have all the files and data-entry structures you need. Create special user records, with a user last name of "Inventory" and locations of "Install," "Stock," and "Repair" for each type of phone. Use the "Quantity" field to record stock levels. Whenever you add, install, pull phones, or send them out for repairs, adjust the levels in the appropriate records, just as you would on paper.

If you want to get fancy, and feel up to the task, you can do some programming to accept transactions and automatically post them to the different inventory records. This has the advantage of enforcing consistency and guaranteeing your records will stay in balance. On the other hand, it's a lot of work to automate a relatively simple chore.

By the Way: Just because you're doing everything else on computer, don't feel obliged to manage your inventory there as well. You can keep your inventory very effectively on paper.

Or. . .use an electronic spreadsheet program for this purpose.

Don't become enslaved by your computer systems. They are tools to help you do your job, and that's all they are. For most of the kinds of recordkeeping described in this book, computerized systems make a lot of sense. For telephone inventory management, they may make sense for you or they may not. Always keep your goal in mind: To know what you have and where it is.

KEEPING TRACK OF CABLE

This is as good a time as any to discuss the "map *vs* schematic" controversy which hasn't raged through the industry. . .yet. (Don't worry, it will. These things take time.)

If you think about cable, an instantly attractive notion is to layout all of your cable runs on a *Computer-Aided Design* (CAD) package. Once you have done this, you can have instant reference to the exact path each cable has taken through the building. This is the so-called "map" approach to cable management, because you're creating a map of your building.

Your alternative: Maintain a schematic diagram of how cables are connected, without diagramming their locations. The schematic theory says that if you know a cable runs from one closet to another, you don't need to know exactly how it gets there.

Opinion: Maintaining a map is too much work, and has too little payoff. You can represent a schematic within the database structures we've already established, so maintaining cable data becomes a matter of recordkeeping rather than drafting.

A further advantage: Since you're keeping your cable data in a computer database, you can search for specific cable records easily, instead of having to peruse a map.

How to Do It: You don't need new database files. Take a gander at the SYSTEMS and PORTS database you've already constructed. It has a field called DEVICEID, for which you should make a separate entry for each wiring closet.

Describe each punchdown block as a CABINET, and each position in the punchdown block as a PORT (ignore the remaining fields in the database).

It's easy to imagine a program that can trace cable from the PBX down to the phone, or vice versa. Just retrieve a record and find the OUTCABLE number. Then, match it to different cords containing the same number as an INCABLE. Repeat the process. That's all.

This isn't a book about programming, but depending on how fancy you want to get, it isn't all that hard to write a program that will print out a complete schematic diagram of your network.

Keeping Track of Work Orders

The key to a successful computerized system is understanding that paper forms are intrusive, while a database is not. In other words, even though you have computerized your work order tracking system, you still need to place a printed work order in your technicians' in-baskets. Counting on them to periodically review their open work orders in the system is simply unrealistic.

You can count on technicians to close out work orders on the computer, but only if you make that an important part of their performance appraisals (if they work for you) or part of contract compliance (if you contract for support). Otherwise, technician mentality says that real work is more important than paperwork.

What a warped perspective.

You can only ask your technicians to close out work orders if they all have their own PCs, and your database resides on a Local Area Network to which they all have access. Don't expect them to go to a central machine which you'll all share for this purpose. It just won't work, especially since the work order system will use other files from your database, such as the USERS file you've already created.

Your work flow, then, begins with filling out work order forms on the computer screen, printing out hard copy work orders, and placing them in your technicians' in-baskets. When they complete an order, they either indicate the completion on paper and place the work order back in your in-basket for you to enter in the system, or they enter the completion themselves.

If you're big on audit controls, you may want one field in the database that indicates their completion, and another that indicates your review. (If you have time for that, you're overstaffed.)

The following typical work order database is far from the last word on the subject. Probably, you will want to tailor it to your requirements and phone system.

Work Order File—File Name = WORKORD

Field Name	Type	Description
WONUMBER	N	Work order number. Assigned sequentially by your computer.

Field Name	Type	Description
CREATOR	A10	The LAN USERID of the individual who entered the work order. Should be filled in automatically by the system.
LASTCHG	A10	The LAN USERID of the individual who last modified the work order record. Should be filled in automatically by the system.
ENTERDAT	D	Date the work order was created/assigned.
DUEDATE	D	Drop-dead date for the work.
ENDUSER	A10	The USERID of the person for whom the work is being performed. Should match a record from the USERS database.
REPAIR	L	True if repair is needed.
REMOVE	L	True if the phone should be removed (the employee has left the company, for example).
NEWCLASS	A10	Class of service to which the user's phone should be changed.
NEWGROUP	A10	Group to which the user's phone should be assigned.
NEWMODEL	A10	Model telephone the user wants installed.
NEWEXT	A4	New extension to be assigned to this telephone and location.
NEWLOC	A30	Location to which the user is moving.
NEWDEPT	A10	Department to which the user is moving.
COMMENT	A90	Miscellaneous explanatory text.
ASSIGNEE	A10	USERID of the technician assigned to this work order.
ENDDATE	D	Date work was completed.
ENDTIME	A5	Time work was completed (format HH:MM).

Figure 17 shows a typical work order data-entry screen. Fields that should be filled in automatically are displayed in italics.

Keeping Track of ACD Stations and Groups

When you purchase an Automated Call Distributor, you almost certainly will buy some form of ACD management system to go with it. Keeping two sets of records on the same subject wastes your time. Use your ACD management system; don't create your own.

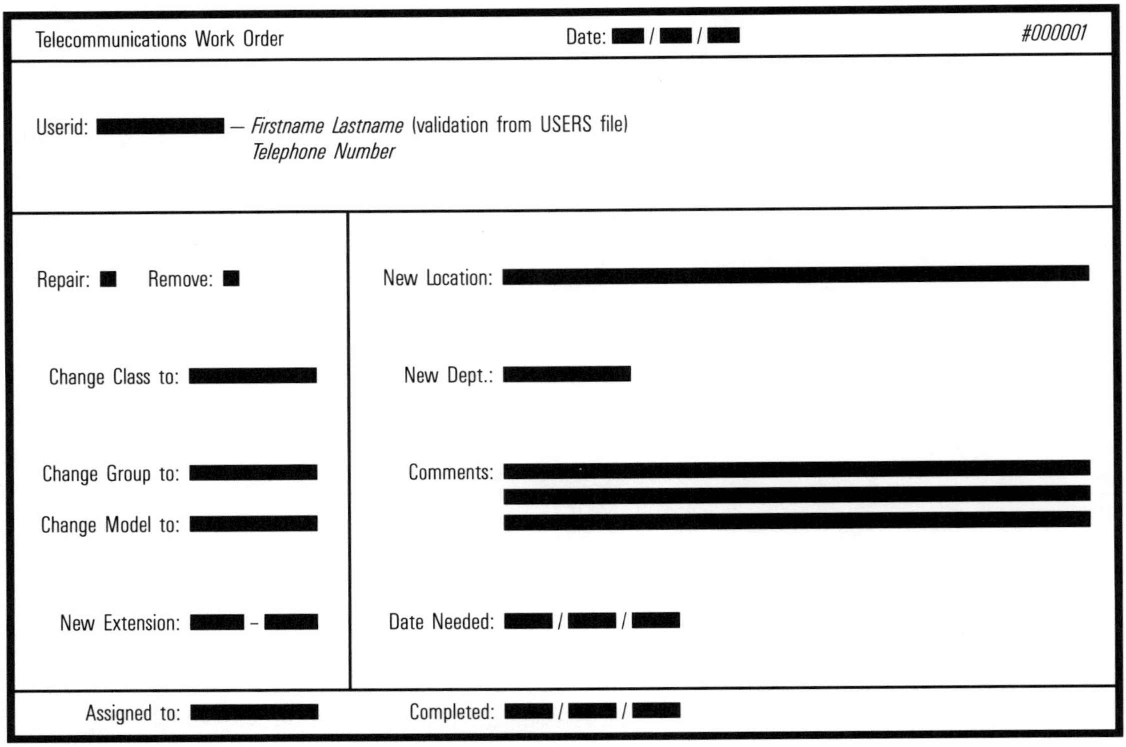

Figure 17.
Work Order Data-Entry Screen

APPENDIX B

Telecommunications Management Forms

THE BOOK HAS PROVIDED A number of form layouts. For your convenience, this appendix contains copies of all of them.

Phone System Ledger Cabinet #_______ Shelf #_______

Card #	Type	Port	Cable #	Ext	Type	User Name	Location

Trunk/Line Ledger Page ________

Trunk/Line #	Type	Carrier	Usage Notes

Telephone Extension Records: From *#5051* to *#5063* Page ______

Ext	User Name/ Cable #	Location/ Closet	Type/ App 2	Cover/ App 3	Group/ App 4	Class/ App 5	App 6
5051							
5052							
5053							
5054							
5055							
5056							
5057							
5058							
5059							
5060							
5061							
5062							
5063							

Telephone Inventory Ledger for Model _________

	Transactions				Inventories			
Date	Add	Install	Pulled	Repair	Install	Stock	Repair	Total

Telephone Cable Ledger for Wiring Closet __________

Cable #	To Closet	To Cable #	Cable #	To Closet	To Cable #

Telecommunications Work Order Date: ___/___/___ # ________

User Name: ____________ ____________ Department: ________

Last Name First Name

Location: ________________________ Contact: ________

________________________ Phone: ________

Phone: ____________

Work Needed:

Date Needed: __/__/__

INTERNAL USE ONLY

Assigned To: ______________

Comments:

Parts Used	
Part	Qty

Completed (Date/Time): Initials:

Telecommunications Work Order Date: ___/___/___ # ________

User Name: ______________ __________________ Department: __________

Last Name First Name

Location: ______________________________ Contact: __________

______________________________ Phone: __________

Phone: ______________

[ ] Repair [ ] Remove
[ ] Chg Class to ______________________________
[ ] Chg Group to ______________________________
[ ] Chg Model to ______________________________
[ ] Other
(attach details)

[ ] New Ext: _____________
[ ] Move — New Location:
___ ______________________________
___ ______________________________

Date Needed: __/__/__

INTERNAL USE ONLY

Assigned To: ________________

Comments:

Parts Used

Part	Qty

Completed (Date/Time): Initials:

Voice Mail User Plan Department #_______________ Page ____

User Name	Ext	Inside Cover	Inside 0-Drop	Outside Cover	Outside 0-Drop	Training Schedule	Done

ACD Group Definition Form					
Group	From	To	Message	Overflow Group	Overflow Condition

ACD Station Assignment Form Group: _______

Station	From	To	Station	From	To	Station	From	To

APPENDIX C

Sample Request for Proposal—Automated Attendant

Table of Contents

I. **Goal:** The XYZ Corporation plans to purchase an Automated Attendant. The system must be able to route approximately 2,000 incoming telephone calls every day, servicing 200 employees. Since the XYZ Corporation anticipates 20 percent annual growth, the system must include provisions for accommodating a corresponding growth in call volume. We request your response to this Request for Proposal (RFP).

The XYZ Corporation, a specialty printing house located in Pittsburgh, Pennsylvania, has been in business since 1953. It caters to small magazine publishers who require low-volume printing runs with short deadlines and high print quality. Our first thirty-five years in business were characterized by slow growth and a stable customer base. In the last five years, demand for the types of service we provide has grown significantly, and our business has grown proportionately.

Our corporate philosophy has been to keep costs reasonable by minimizing administrative overhead, while providing the maximum in service to our customers. The automated attendant fills a key role helping us to achieve this goal; we hope its implementation will provide an alternative to increasing staff.

As our business increases, we have also gained customers from outside our geographic area, increasing the number of calls received after our normal business hours. We plan to use our automated attendant to process these calls.

II. **General Instructions:** All responses to this RFP must address each defined requirement, and should follow the same outline format and numbering scheme, to assist us in evaluating responses. Responses must be received no later than March 15, 1993.

Send four copies of your response to:

Jan Peterson
Telecommunications Manager
XYZ Corporation
123 Main Street
Pittsburgh, PA 12345

Any deviation from the specifications described in this RFP must be explicitly noted and explained. Please feel free to suggest additional technology or approaches beyond those specified in this document. All bids must be good for ninety days. If you have any questions regarding this RFP, direct them to Jan Peterson, 717/555-5555.

Since our field has become heavily competitive, we ask that you treat all information in this RFP as confidential.

III. **Requirements:** We view the following as our major requirements. In your responses, you are encouraged to make recommendations that exceed the requirements as we have defined them.

A. *General Requirements:* We require all equipment to be new, not refurbished. All components must be readily-available items—we prefer to purchase no custom components. The recommended model must be based on current, but mature technology. We expect it to operate reliably for at least five years, and require a guarantee that replacement parts and software upgrades be available for that period of time.

In your response, please specify all components, quantities, unit prices and total prices, and explain the purpose of each component. This information will allow us to explore variations on the basic scenario described in this RFP.

B. *Technical Specifications*

1. *Reliability:* Describe the factors you have engineered into your automated attendant to ensure reliability, including redundant components, built-in diagnostic routines, quality assurance procedures used during manufacture, and any other aspects of your system that contribute to minimum downtime.
2. *Stage in Life Cycle:* Indicate how long this product has been commercially available, and how long your products typically stay on the market. Our preference is to purchase technology that has been available for at least twelve months, but not more than thirty months.
3. *Environment:* Please specify your product's environmental requirements. Include the range of temperatures and humidities within which it operates reliably and any requirements for additional air conditioning beyond basic office loads. If raised flooring will be required, please indicate this.

   Also describe the space needed to house your system, including additional clearance around the unit needed for access, along with the total system weight and floor loading.

   Describe your system's electrical requirements. If your system includes provisions for surge protection and/or an UPS, please indicate this. Otherwise, please provide specifications and pricing for these as options in your proposal.

4. *Features:* Please provide a description of the following features as they pertain to your system:
   a. *Length of message:* Please describe the restrictions your system places on greeting length.
   b. *Number of messages:* Describe the number of different greetings your system can store.
   c. *Time-of-day message activation:* Indicate whether your system can automatically switch between different greetings at different times of day. E.g. one message for business hours, the other for after-hours telephone answering.
   d. *Number of autoattendants:* Please indicate whether your system can provide more than one automated attendant, to handle different incoming lines. Our initial use will be to automate a single, main telephone number, but in the future we may wish to publish separate numbers for sales, customer service, and general business calls.

      If the use of multiple attendants imposes any additional constraints in message length or number of messages beyond those already described, please indicate the nature of those constraints.

      Please describe how we would expand your system beyond its current capacity, should we develop additional uses for automated attendants beyond the capacity we are initially purchasing.
   e. *Menu processing:* The system we select must allow us to present a menu of choices, selected by touchtone. Please indicate the number of menu layers your system allows. If your system allows multiple greetings and automated attendants, indicate whether each supports its own set of menus, and what constraints the system imposes on implementing multiple menus associated both with different greetings activated by time-of-day, and with multiple attendants supporting different incoming lines.

Describe the provisions for system expansion we have available to us to grow beyond the constraints you have just described.

f. *Zero drop:* The system must allow callers to press 0 at any time and immediately switch their call to a live operator. If your system supports multiple greetings and attendants, each must allow definition of its own zero drop destination.

g. *Rotary phone handling:* We would prefer to purchase a system that can interpret rotary phone pulses as well as touchtones. We anticipate expansion into the "Vanity Press" business; many potential customers live in rural areas still dominated by rotary telephones. We are open to third-party rotary-to-DTMF convertors, so long as your organization will take responsibility for selecting, installing, and maintaining the equipment. If any other alternatives exist for handling calls from rotary telephones, please describe them.

h. *Voice editing:* Please describe the provisions your system makes for editing announcements. Explain whether changes in announcements result in re-recording that announcement in its entirety, or if you allow editing and partial re-recording.

5. *Programming and Administration:* Describe the process of setting up and keeping track of greetings/announcements, menu trees, zero drops, and so on, along with the tools your system provides for managing these once they have been recorded and programmed. Also describe the training required in order to become competent in ongoing system programming and management.

   Please indicate whether your organization will contract to provide ongoing programming and administration for our system, as an alternative to our providing in-house support. Include in your proposal pricing options for this service.

6. *Backup and Recovery:* Describe the provisions your system makes for backup and recovery. In case of different levels of system failure, we prefer to avoid re-recording announcements and reprogramming

menus. Indicate whether your system backs up announcements, menus, or both, and whether it provides options for selective restoration of menus and announcements, if your system supports multiple greetings.

7. *Integration:* Your system will have to integrate with a Rolm 8000 PBX supporting one main attendant console and 200 telephones. Please describe *in detail* how your system interfaces with the Rolm 8000. Include both a block diagram and a narrative describing the interface. Include explanations of how your system receives and transfers calls, and whether we will have to reprogram the PBX to accommodate your system.

   We are considering replacement of our Rolm system with a more modern PBX or hybrid system. Please indicate any limitations as to the systems to which your automated attendant can interface.

C. *Business Specifications*

1. *Vendor Experience:* Please indicate how many systems your organization has installed, specifically of the model you are recommending in this proposal. Please indicate how many of these you have installed in the state of Pennsylvania, and how many are within fifty miles of Pittsburgh.
2. *Warranty and Maintenance Terms:* Provide a copy of your standard warranty, and an explanation of any unusual terms or conditions.

   We will want to contract for maintenance rather than paying time-and-materials charges. Provide a copy of your standard maintenance contract, and provide pricing as part of your pricing proposal. If you have a provision for after-hours coverage, please provide information and pricing for this option.

   Please indicate how many service technicians you have located within the Pittsburgh metropolitan area, and how many other systems they currently support. Only include technicians trained in the system you recommend in this proposal.

   If your system includes provision for remote

maintenance by modem, include a description of this option, including the types of system failure for which you can provide remote diagnosis and/or recovery.

3. *Payment Terms and Conditions:* Indicate payment terms and conditions as part of your pricing proposal.
4. *References:* Please provide three references that have bought the same product you recommend in this proposal.

Index

Y

Z